COASTAL
EROSION

OTHER BOOKS BY ROBERT FRITCHEY

WETLAND RIDERS

THE GULF WARS SERIES:
MID-1990S FIGHTS OVER NETS
ON THE GULF OF MEXICO

MISSING REDFISH

LET THE GOOD TIMES ROLL
(Louisiana)

A DIFFERENT BREED OF CAT
(Alabama)

COASTAL EROSION

Mississippi Forces Commercial Fishermen to Work with Unworkable Nets in Impossible Locations

ROBERT FRITCHEY

N E W
MOON

HARRISBURG, PENNSYLVANIA

Published by New Moon Press

www.newmoonpress.com

Edited by Jerry Fraser
Book and Cover Design by Kathleen Joffrion
Line drawings by Lee Roy Tooke
Photos are by the author unless noted otherwise.

ISBN 978-1-7346171-9-1

Library of Congress Control Number: 2025904816
Printed in the United States of America

For information regarding special discounts for bulk purchases,
please contact:
www.newmoonpress.com

For tenacity

PREFACE TO THE GULF WARS SERIES:
MID-1990S FIGHTS OVER NETS
ON THE GULF OF MEXICO

The Gulf of Mexico would be landlocked but for the Yucatan Channel and Florida Straits. Warm Caribbean currents flood into this basin through the channel and the overflow jets out the straits and up the Atlantic Coast as the Gulf Stream.

The Gulf is also fed by rivers that enter from Mexico and the five American states on its northern shoreline. Rich in nutrients, their fresh inflow lingers in indentations along the coast, and barrier islands further impede mixing with the salty sea. Such protected estuaries, with their softened blend of fresh and salt water, serve as nurseries for an abundance of seafood species.

Oysters, clams, crustaceans such as shrimp and crabs, and a myriad of finfish thrive there.

Native Americans naturally harvested this bounty, and coastal tribes had little need to venture inland. Later, the Spanish and then the French colonized parts of the coast, giving the Gulf a Mediterranean flavor that was soon diluted by a steady influx of English and Scotch-Irish from the American states.

While the earliest settlers harvested seafood for their own subsistence, by the late 19th century the American Gulf Coast from Key West, Florida, to Brownsville, Texas, was dotted with coastal communities whose economies depended almost exclusively on renewable marine resources. Such centers of commerce attracted successive waves of workers from around the world including the Canadian Maritimes,

China, the Philippines, the Canary Islands, Italy, Greece, Scandinavia, Croatia, Vietnam, Cambodia and Mexico.

Many of these immigrants came from fishing communities and brought harvesting and processing techniques that advanced the development of the Gulf's fishing economy.

In the earliest days, all the fishing necessarily occurred near shore. By the 1880s, a fleet of Pensacola, Florida-based schooners was sailing to the Campeche Bank, off Mexico's Yucatan Peninsula, to handline red snapper. From such beginnings, the offshore industry evolved into that of today, with far-ranging diesel-powered vessels that bring in tuna, swordfish, grouper and other deep-water fish, and steel-hulled shrimp trawlers massive enough to be called "slabs."

Through it all, the smaller-scaled artisanal fleet continued to develop in the shallows along the Gulf's edge. From nets of treated cotton or linen, and boats driven by wind or muscle, the inshore fishermen progressed to speedy inboard- or outboard-powered skiffs, and webbing of modern synthetics. Likewise, the buyers and sellers of fish continually probed new markets for the fishermen's catch, and in so doing nourished seafood consumers locally, nationally, and internationally while steadily bolstering this sustainable industry's contribution to local, state, and national economies.

As the 20th century drew to a close, a convergence of forces virtually eliminated this industry and the way of life it supported.

Wild seafood species multiply free of charge, without planting, fertilizing, irrigating, weeding, or spraying. The annual crops need only to be harvested yet the waters don't willingly yield their bounty, and the arduous work leaves little in reserve for the type of tending that the fish do need.

PREFACE TO THE GULF WARS SERIES

To sustain their abundance, coastal fish require clean, free-flowing waters and lush natural habitats. These simple needs are threatened by a plague of societal impacts. And as society grows, so do the threats.

Unable to lobby or vote, the fish are best represented by those who depend most directly upon them. It's an easy fit for commercial fishermen because, as unabashed natural predators, fishermen require the same pure waters and unblemished landscapes as their prey.

Yet, even while the fishing industry grew steadily in value, it was outstripped in economic and political influence by newer coastal enterprises: From the western Gulf, spreading east, there was the oil business, and from the eastern Gulf, spreading west, tourism.

Like fishermen, oilmen labor to produce a unique product, export it from the coast, and bring in money. But without the limits inherent to sustainable endeavors, the mining of fossil fuels—mostly in Texas and Louisiana—has mushroomed to such a gargantuan scale that it has miniaturized commercial fishing's political and economic importance. Meanwhile, the natural incompatibility of oil and water has weighted fish and fishermen with additional stresses from pollution and habitat destruction.

For tourism boosters, the Gulf Coast has proved an easy sell, though west of Florida sandy playgrounds—beaches—are limited. The Gulf's primary selling points are its temperate winter climate and the same abundant resources that built the commercial fishery.

Recreational angling began to take off during the plush Gilded Age of the late 19th and early 20th centuries. So did the production of related fishing gear, books and magazines. Tracking the growth in the nation's population and economy, the sport—and the industry supporting it—suffered downturns during World War I and the Great Depression, grew steadily during the 1930s and began to boom amid post-World War II prosperity, as evidenced by the explosion in fishing tackle sales, from $35 million in 1939 to $130 million in 1947.

From the 1960s through the 1990s, as the nation's 78 million Baby Boomers came of age, the number of sport fishermen shot through the roof.

Commercial fishermen may not have welcomed the newcomers with open arms, but they did not begrudge them their fair share of the resource. The same cannot be said for the sportsmen, who covetously eyed the seafood harvesters' catch and needed little prodding by recreational media and other industry members to lobby for increasing shares of this public resource; as they did, the sportsmen consistently portrayed the traditional food providers as pillagers and themselves as "conservationists," an effective yet cynical strategy considering that, contrary to public perception, the recreationists' cumulative haul had in many cases grown to far exceed that of the commercial harvesters.

Nonetheless, the sportsmen's chorus was mysteriously amplified in the early 1990s when virtually every major environmental non-profit in the country joined in an unprecedented environmental education campaign that left the general public convinced that every fish everywhere was endangered by commercial fishing.

With that advantage, and in an orchestrated effort, sport-fishing interests around the country mobilized to ban the use of the most ancient and essential of the coastal fishermen's harvesting tools—the net.

Away from the Gulf of Mexico, commercial fishermen and their allies for the most part rebuffed the sportsmen's efforts. The distant fights—in Alaska, Oregon, Washington, Pennsylvania, New Jersey, and North Carolina—all revolved around some coveted trophy fish, namely salmon, walleye, or striped bass.

PREFACE TO THE GULF WARS SERIES

On the fertile Gulf it was the wetlands-loving redfish that stirred the passions of coastal anglers and led to the seminal 1980s fish fights in Texas, where sportsmen rallied behind elite leaders to monopolize the species as a "gamefish," then went on to banish nets from that oil state's waters. Encouraged by those victories, and in collaboration with the same political influencers, like-minded anglers across the Gulf would wage their own campaigns, which culminated in the mid-1990s net-ban battles in Florida, Alabama, Mississippi, and Louisiana.

Recreational fishing advocates made outlandish claims. I recognized them because I'd supported myself for a time by fishing with nets. So, with a passable ability to craft sentences, I took on the task of writing a book that would help to set the record straight. In time, it became apparent that the completion of such a comprehensive state-by-state effort would be too far in the future, so I broke it down into a collection of separate books, the "Gulf Wars Series."

The first—"Missing Redfish"—is a management history of the Gulf's most contested fish, which commercial fishermen targeted primarily with nets.

The second book in the series, "Let the Good Times Roll," chronicles the battle over nets—and redfish—in Louisiana, a state that calls itself the "Sportsman's Paradise." In "A Different Breed of Cat" Alabamians grapple with the issue.

"Coastal Erosion" recounts the death-by-a-thousand-cuts that was delivered upon Mississippi's band of netters.

Net fishing is a folk occupation, handed down over the generations from fisherman to apprentice, so banning the use of nets is a one-way train that virtually eliminates not only the ability to produce seafood but a traditional and sustainable way of life. These wars on the Gulf, therefore, marked the end of an age and the beginning of another in every state.

Some of the changes associated with this transition can be readily quantified, such as the plummeting commercial landings of wild-caught fish.

As for the long-term effects on the actual health of the Gulf Coast's environment and fishery resources, only time will tell. One certainty is that those resources will be worth far less to society; another is that their actual status will be more difficult to assess minus the ongoing interaction of professional harvesters.

Because losses to our coastal heritage are even more difficult to measure, I've supplemented the historical narrative of each state's battle with some in-their-own-words profiles of fisherfolk whose livelihoods depended on fish.

Even though I fished for years, commercially and recreationally, I learned much in writing this series—not just about the Gulf of Mexico's ecology and fisheries, but how mass movements are organized, and how man relates to himself, his fellows, and to nature. It's often not a pretty picture.

"Sustainable use" is simply the practice of continuing to do what commercial fishermen, as active participants in the natural world, have always strived to do—use renewable resources in a way that doesn't deplete them or harm the environment. According to its proponents, if you don't use living resources you lose them, mostly by reducing their value to society and diminishing incentives for their stewardship.

As we occupy the driver's seat of our own evolution that's worth considering. Otherwise, our ready tendency to believe in the paucity of those resources may prove self-fulfilling.

—Robert Fritchey, 2025

CONTENTS

CONTENTS, CONTINUED

CONTENTS, CONTINUED

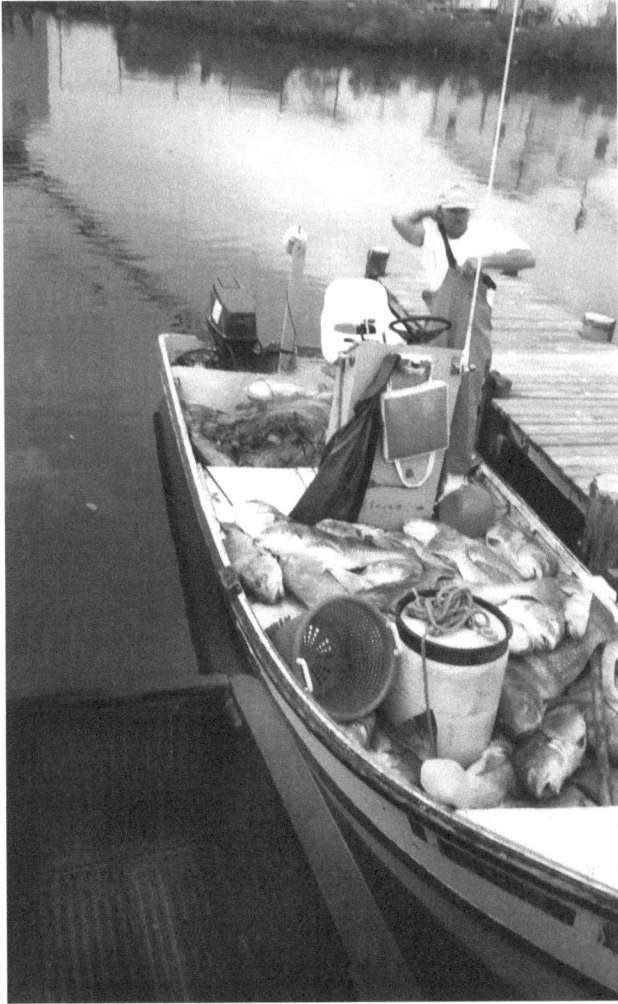

Pascagoula net fisherman Hilton Floyd with a catch of "bull" black drum in the early 1990s. *(Courtesy of Hilton Floyd)*

INTRODUCTION

"Coastal Erosion" is in two parts: "The First Contagion" recounts the battle in Mississippi during the recreational fishing industry's mid-1990s attack on virtually every commercial net fishery in the country. In the second part, the survivors are whittled down in the 21st century to a single old-time haul seiner— the "Last Man Standing."

As in virtually every other state at the time, the fish fights in Mississippi weren't violent, they were political. And in political campaigns there's not a lot of flattery. To win, sportsmen relentlessly demonized netting until anyone might have expected fish populations to bloom as soon as the dreaded gear was pulled from the water.

Sustaining fishery abundance over the long term, however, hasn't been that simple. The state's coastal fish faced a variety of real threats—some natural, some manmade—that needed to be addressed on an ongoing basis, forever.

They included habitat loss, pollution, and, yes, removals by both sport and commercial fishermen.

HABITAT LOSS

Mississippi lies within the "Fertile Fisheries Crescent." The term was coined in the 1960s by Gordon Gunter, Ph.D., the director of the University of Southern Mississippi's Gulf Coast Research Laboratory, to describe the immensely productive central-Gulf waters between Alabama's Mobile Bay and Sabine Pass on the Texas/Louisiana border. Gunter attributed the region's enhanced fishery production to its expansive tidal marshes and massive inflow of fresh water from several rivers including the largest in North America—the Mississippi—which empties into the Gulf almost directly south of Biloxi.

The marshes formed the base of the coastal food web, and populations of seafood species were directly proportional to the area of those rare wetlands. By the 1990s, it was common knowledge that the marshlands within the fertile crescent were diminishing: Louisiana led the nation in the amount it was losing, and Mississippi was second.

The Magnolia State had been losing an annual average of 200 acres of wetlands per year since the 1850s; ten thousand acres vanished in the sixty years from 1947 to 2007. Natural subsidence and erosion from wave action took some, but most of that loss was related to development.

Wetlands withered behind seawalls that were built to stabilize shorelines and were even more directly impacted when they were converted to dry land.

A 2013 report by Mississippi's Department of Marine Resources concluded that, between 1930 and 1973, nearly 8,200 acres of coastal marsh were drained and filled for industrial and residential use. Between 1972 and 2000, wetland acreage in the state's three coastal counties decreased by one third.

Mississippi's coastal land loss wasn't limited to its wetlands.

A shoal crossed the state about ten miles off its shoreline and delineated the inshore estuary from the open Gulf of Mexico. The barrier emerged in several spots as islands with churning sandy beaches on their seaward edges and meadows of submerged seagrasses in their calmer lee waters. Not only did the barrier islands expand and diversify fishery habitats in their immediate vicinity, they helped regulate the salinities within the entire estuary.

Most of those islands were shrinking: The total amount of open water between them has increased by fifty percent since 1850, from about

23 miles to 33 miles in 2018.

Continued erosion of those protective islands, along with rising sea levels, presaged an increasing influx of salty Gulf waters to the nursery. That could hurt the seafood species that in their early stages typically favored sweeter water. Then again, some dilution from the sea could help protect those same fisheries from threats that might arrive with the inflow of fresh water from inland.

POLLUTION

Horn Island is the longest of the state's barrier islands. It stretches ten miles in an east-west direction, and underwater seagrasses proliferate behind it.

Like marshes, seagrasses provide food and shelter for seafood species. The more the merrier. It was therefore gravely concerning when, in the 1980s, those grass beds suddenly began to die off.

Because Horn Island lies directly south of the Pascagoula River's mouth, when a local botanist tried to identify the cause for the decline, he looked upstream.

Lionel Eleuterius, Ph.D., specialized in wetland and marine plants and was hired in the 1960s by Gordon Gunter, at the Gulf Coast Research Lab in Ocean Springs. In 1986, he reported that two of the three major species of Horn Island's seagrasses were completely wiped out. Eleuterius suspected that the decline was related to the discharge of dioxin—a toxic by-product of the paper bleaching process—from a pulp plant that had recently opened about 80 miles upriver.

The $560 million Leaf River Forest Products wood-pulp mill began operating on the Leaf River near New Augusta in 1984. The Leaf River merged downstream with the Chickasawhay River to form the Pascagoula River.

Although the scientist was unable to obtain funding to pinpoint the reason for the seagrass decline, the circumstantial evidence was damning. Dioxin, a component in Agent Orange—the herbicide and defoliant used by the United States in the Vietnam War—was found in the wastewater and sludge of the mill.

The chemical was later found in the tissues of freshwater fish which prompted the Mississippi Department of Wildlife, Fisheries and Parks to issue consumption advisories for fish caught from the Leaf and Pascagoula Rivers, and to temporarily close those rivers to commercial fishing in 1990.

The release of pollutants from an identifiable origin—"point-source" pollution—wasn't as common as it used to be thanks to government regulations, costly liability, an evolving economy, and a generally heightened environmental awareness among the public. "Non-point-source" pollution was more insidious.

Rainwater washed nutrients, pesticides, and herbicides into water-bodies from agricultural operations, and picked up a cocktail of pollutants as it ran over impervious surfaces such as roofs, streets, parking lots, even lawns, associated with residential and industrial development.

It was a race—as destruction of the natural environment snowballed, it became increasingly difficult—politically and economically—to resist further development. Zoning, stormwater management, and other building regulations helped, but the single most effective way to protect coastal fisheries—and virtually all other wildlife—from both pollution and habitat loss was to keep as much land as possible in its natural state. And the surest way to do that was to buy it.

"Public ownership in fee simple (the absolute ownership of land with unrestricted rights of disposition) is really the ultimate way to protect our estuaries and coastal wetlands," stated fishery biologist David Ruple in 1997.

INTRODUCTION

Ruple was the director of the Mississippi DMR's Coastal Ecology Division and had been instrumental in developing the state's Coastal Preserves Program in 1992.

The program's objective was to "acquire, protect and manage" sensitive coastal habitats and, by 1997, over 21,000 acres had been set aside. More acquisitions would follow, but in the race between development and preservation, development got a jumpstart after the Mississippi Legislature legalized dockside gambling in 1990.

The Gulf Coast became the fastest growing region in the state. From 1990 to 2000 the number of visitors increased nearly eight-fold, from 1.5 million to 11.5 million. The tourists wanted to gamble, sunbathe and swim, play golf, dine on seafood, and go fishing.

FISHING

In the 1990s, more people were going sport fishing more often, not just in Mississippi but across the nation. So, the local surge wasn't unique. But on a coast that spanned just seventy miles, it was concentrated.

The federal National Marine Fisheries Service estimated that recreational anglers made 2.6 million trips in Mississippi's coastal waters in 1990. By 2000, that number had grown nearly 70 percent, to 3.8 million.

By 1994, when Magnolia State sportsmen began to press for their ban on gillnets and trammel nets, they—like saltwater anglers in the other Gulf states—were already hooking the lion's share of the popular coastal species that the seafood producers were netting for consumers. And biologists at the state's Department of Marine Resources weren't shy about stating that fact.

Normally, that sort of information would hinder an interest group that was trying to grab an even greater share of the resource. But the mid-1990s were anything but normal.

5

Newspaper headlines from late 1993 to early 1994.

INTRODUCTION

Across the country, objective scientific data were being buried beneath a blizzard of fish propaganda that was unprecedented in its quantity and quality. The spare-no-expense media campaign of the Pew Charitable Trusts, with its history of antipathy toward commercial fishing, and other "nonpartisan" philanthropies, along with publisher Times Mirror Magazines and other outdoor press, rhetorically carpet bombed the public until "overfishing" and commercial fishing became reflexively linked.

Meanwhile, recreational industry leaders ginned up a movement that anglers were only too willing to join. Neither knowledge nor experience in fishery management was required to join, just the easy belief that no one should be allowed to catch more fish than they were.

Fully aware that they'd been handed a once-in-a-lifetime opportunity, Mississippi's activist anglers attacked with a vehemence that stunned fishery managers, some of whom, as chance had it, weren't too experienced themselves.

Working fishermen fought the ban as best they could. But they were outnumbered, short on allies, and worn down after years of fighting to maintain access to the Magnolia State's coastal waters.

In the end, their gear wasn't "banned," they were simply forced to work with unworkable nets in impossible locations.

After the 1990s battles, resourceful netters probed for a way to get back fishing. Around 2015, some of them exploited a loophole in the regulations that allowed them to make some good catches.

But the workaround wasn't sustainable, and it stirred up a controversy that revealed an old-time net fisherman who'd been legally working under the radar for years. As this book went to press in 2025, he was still at it— the last man standing.

PART ONE

THE FIRST CONTAGION

Populations and landings of red drum, speckled trout, and mullet, plus the number of net fishermen over the past 20 to 30 years, have remained stable. Looking at the landings data and fisheries data, we have seen no downward turn.

—Tom Van Devender, fishery biologist,
Mississippi Bureau of Marine Resources, 1995

With gillnets out there, we are going to overfish the Gulf.

—Ray Lenaz, state president,
Coastal Conservation Association of Mississippi, 1995

The crisis they are portraying does not exist. Commercial fishermen are taking a select part of a renewable resource.

—Pete Floyd, commercial fisherman,
Save America's Seafood Industry, 1995

In my view, there was no evidence to suggest a total gillnet ban was warranted. The gear is very selective and can be managed by mesh size, length and time or location of set. And even though sportfishermen complain that it's too efficient, efficiency is exactly what you want. We want efficient cars, why wouldn't we want efficient nets that do the job they were designed to do? Asking gillnetters to quit gillnetting is like asking sportsmen to fish from row boats. There's no reason to impose artificial handicaps if the science doesn't show it's needed.

—Vernon Asper, Commissioner,
Mississippi Commission on Marine Resources, 1997

CHAPTER ONE

DEATH BY A THOUSAND HEARINGS

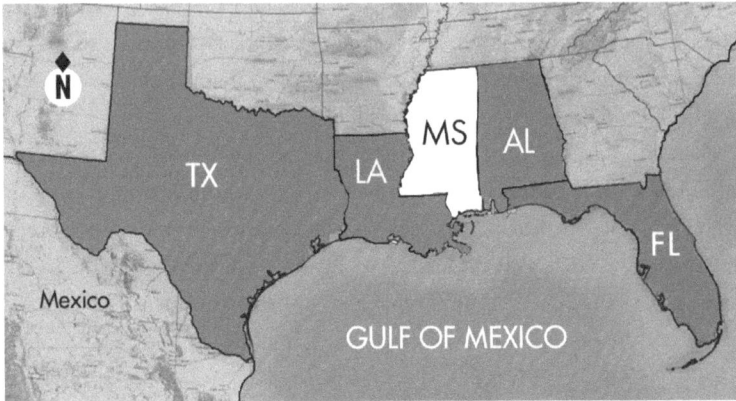

Like the wings of a butterfly, the central-Gulf states Mississippi and Alabama were mirror images of each other—they both extended inland to the Tennessee line, had about the same limited frontage on the Gulf of Mexico, and supported outsized marine fisheries that were generated primarily within a single hyper-productive body of water.

While Alabama's Mobile Bay appeared as a deep inland-reaching cleft that bisected the state's coast, the Mississippi Sound spanned the Magnolia State's entire coastline and extended as far east as Mobile Bay and as far west as Louisiana's Lake Borgne.

A comparatively shallow basin, less than 20 feet deep, the Sound was delineated from the open Gulf of Mexico by a chain of barrier islands staggered across the coast about 10 to 15 miles offshore. The broken ridge of islands and bars lessened the wave action of the Gulf and helped retain the fresh water and nutrients that flowed into this half-million-acre estuary from the region's inland waterways.

From east to west across the coastline, four rivers emptied into the Sound—the Pascagoula, Biloxi, Jourdan, and Pearl. The coastline itself was scalloped with the Pascagoula, St. Louis, and Biloxi Bays, each with the full complement of salt marsh, grassy islands and bars, oyster reefs, and underwater meadows of aquatic vegetation that generated estuarine sea life like crabs, shrimp, and finfish.

Like the state of Alabama, which protected the marshy upper reaches of Mobile Bay, Mississippi maximized its seafood production by managing its inshore nurseries as sanctuaries where commercial harvesting, at least, was off limits. When they matured and grew into a marketable size, the shrimp, crabs, and finfish fell out into the deeper waters of the Sound where they become fair game.

RECREATIONAL FISHERIES

Mississippi's Legislature first authorized a saltwater sport-fishing license in 1993. In 1994, out of a statewide population of 2.5 million, about 56,000 residents bought licenses. An additional 44,000 or so anglers were exempted from buying the new license because they were either disabled, younger than 16, or older than 65.

So, an estimated 100,000 sportsmen and sportswomen wet their lines in the state's coastal waters in 1994. According to the federal National Marine Fisheries Service, those anglers made more than 2.7 million saltwater fishing trips, an 85 percent increase over the 1.5 million trips taken in 1981, the first year the agency began to compile estimates of sport-fishing effort.

Some of those trips were for offshore species like snapper and tuna but most recreational anglers fished nearer to shore where a variety of tasty species was accessible.

In 1994, sport fishermen in Mississippi caught over 4.4 million

The Mississippi Sound spans the entire coast of Mississippi by extending east to Alabama's Mobile Bay and west to Louisiana's Lake Borgne. Waters within this estuary are about half as salty as those in the open Gulf.

pounds of their 14 most popular species of inshore fish.

In descending order of volume, those landings included sheepshead (1,584,243 pounds); sand seatrout (606,276 pounds); red drum (448,295 pounds); black drum (410,695 pounds); spotted seatrout (404,698 pounds); southern and Gulf flounders (292,251 pounds); striped mullet (287,282 pounds); southern and Gulf kingfish (164,274 pounds); croaker (136,731 pounds); Spanish mackerel (72,433 pounds); pinfish (41,164 pounds); and gafftopsail catfish (4,187 pounds).

These were essentially the same inshore species that the state's commercial fishermen targeted with their nets.

COMMERCIAL FISHERIES

Fishery landings are as variable as the Dow Jones Industrial Average but because 1994 was the last year that catches were unimpacted by the net-ban controversy, it will serve as our benchmark.

In 1994, Magnolia State commercial fishermen landed a total of 220.2 million pounds of saltwater products, with a dockside value of nearly $37 million.

As usual, menhaden purse seiners accounted for around 90 percent of the total commercial haul, or 195.5 million pounds with an ex-vessel (before processing) value of $9.2 million.

Menhaden are too oily for most people's taste, but the small herring has many industrial uses: Some are frozen and marketed as bait but most of the catch is reduced to meal that is fed to poultry and livestock, and oil that is used to manufacture paints, plastics, resins, even omega-3 supplements.

Mississippi's shrimp trawlers usually caught at least 10 million pounds of brown, white, and pink shrimp each year; in 1994 they brought in a little less—8.7 million pounds worth over $22 million at the dock.

(To capture the full economic impact of seafood products as they ascend the market ladder to the consumer, dockside values are multiplied at least 3.5 times.)

From 1950 to 1994, Mississippi's oyster landings ranged from a high of 4.8 million pounds in 1964 to a low of 21,000 pounds in 1980. In 1994, oystermen produced a respectable 1.7 million pounds of "meats" worth more than $2.3 million.

Annual landings of blue crabs between 1950 and 1987 averaged well over 1.5 million pounds, reached a high of over 4 million pounds in

(*Opposite*) Menhaden fishermen hardening up their catch in a purse seine before pumping it aboard the mother ship. The low-value industrial species must be produced in great volume to be profitable for companies such as Pascagoula's Omega Protein corporation. Conversely, more valuable edible species like redfish, trout, and mullet could be harvested by just one or two independent fishermen with a gillnet or trammel net.
(*Brian Gauvin photo*)

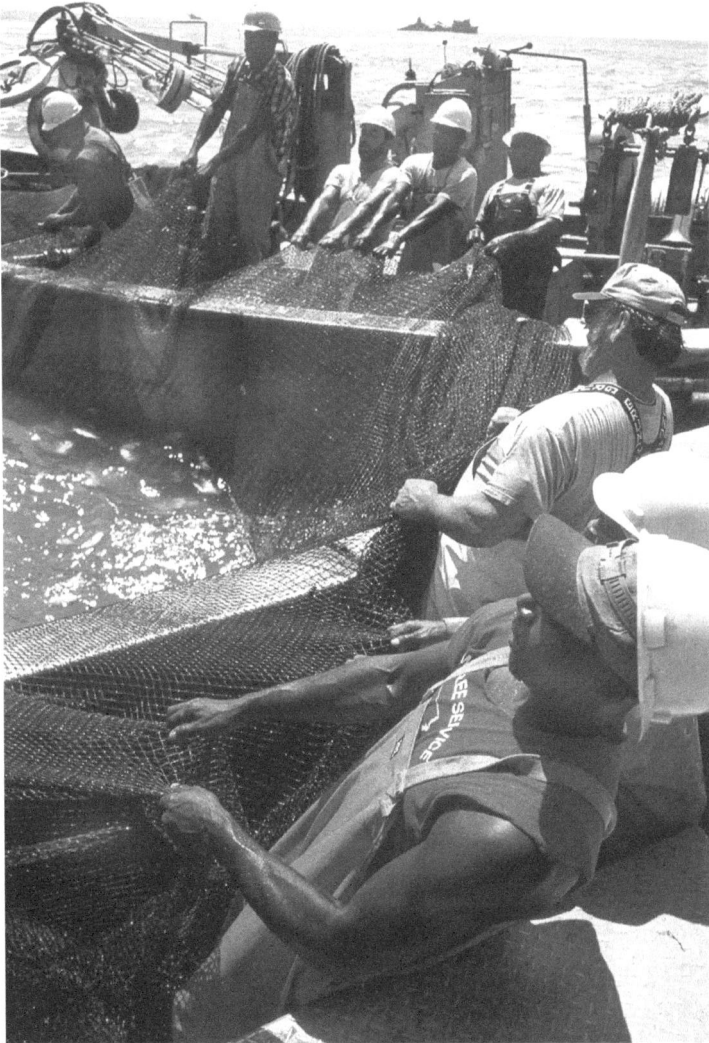

1950, and dropped below 1 million pounds only once, in 1962 when fishermen trapped nearly 910,000 pounds. Crab landings declined after 1987: From 1988 through 1994 they averaged less than 465,000 pounds per year, and in 1994 fell to a 44-year low of 172,000 pounds worth $92,000.

(According to the Gulf States Marine Fisheries Commission's 2001 Gulf of Mexico Blue Crab Fishery Management Plan, the sustained reduction in landings was attributable to social, economic and regulatory changes in the fishery, not to real declines in the crab population.)

Landings of edible saltwater finfish in 1994 totaled 14 million pounds, which included offshore species such as red snapper (151,362 pounds worth $278,195) and more than 12 million pounds of unclassified species that were of comparatively low value, at $1.7 million.

Landings of the 16 edible species that were caught primarily in nets totaled nearly 1.5 million pounds, earning fishermen $860,000.

In descending order of volume, net-caught species in 1994 included striped mullet (781,300 pounds, $407,103); sheepshead (228,910 pounds, $46,000); sand seatrout (109,823 pounds, $59,979); king whiting (78,614 pounds, $35,866); spotted seatrout (73,034 pounds, $123,223); black drum (56,853 pounds, $30,854); flounder (40,753 pounds, $56,205); red drum (40,246 pounds, $48,927); Spanish mackerel (37,520 pounds, $13,298); spot (12,687 pounds, $3,057); Florida pompano (9,548 pounds, $33,060); bluefish (3,220 pounds, $773); croaker (2,447 pounds, $789); ladyfish (2,155 pounds, $759); sea catfishes (1,668 pounds, $321), and tripletail (709 pounds, $539).

From 1990 through 1995 Mississippi issued fewer than 200 commercial net licenses per year.

The fishing area off the 70-mile coastline was comparatively limited to begin with and reduced further by seasonal and spatial netting closures that were enacted to accommodate recreational anglers.

So, unlike in Florida or Louisiana, it was difficult to earn a living in Mississippi by exclusively netting saltwater fish. Most license holders were therefore part-timers who either held jobs in fields unrelated to fishing or were full-time commercial fishermen who engaged in other fisheries until the finfishing picked up.

Bob Metz, on Bayou Caddy, just six miles from the Louisiana line, was one of the latter: "I've got a 23-foot mullet boat with a 150 [horse-power] Suzuki, I've got a 17-foot cheap rig—I've got five boats, two diesel-powered. I'm boat poor," said Metz who crabbed year-round, and "when the fishin' gets good," netted species like spotted seatrout, which came ashore in the spring, and mullet, which migrated offshore in the fall to spawn.

"Mullet's our biggest fishery," he said. "I don't miss a day from about the fifteenth of October until about the first of December. I'm out in rough, rain, whatever, I fish all day every day and sometimes at night too," said Metz, who—just after Florida's November 8, 1994, election—summed up the onset of the campaign to end such work: "We were all fishin' mullet when it started. We knew from the papers and from talkin' to people that they were gonna make an extra big push after Florida. But not like this."

HERE THEY COME

Mississippi's activist anglers had been closely tracking the progress of the Florida Conservation Association's ballot measure. So, when the vote went their way in November, they were ready to go.

Marshalled by the Texas-based Coastal Conservation Association (also known on the Gulf as both the CCA or GCCA, an abbreviation of the group's original name, the Gulf Coast Conservation Association) and coordinating with its local leaders in the other central-Gulf states, the sportsmen would propagandize in the media and pressure every coastal city council and county governing board to take a stand on nets. They'd get a boost from the national sport-fishing industry, the state chapter of a mainstream environmental group, and fellow travelers in the press. Even the National Park Service would pile on the commercial fishermen, as would the state's legacy wildlife and fisheries agency.

A NEW DAY

The Mississippi Department of Wildlife, Fisheries and Parks was overseen by a commission of five members, each appointed by the governor from one of the state's five congressional districts. Since all three of Mississippi's coastal counties were located within the same district, the panel couldn't include more than one coastal representative, who was by default outnumbered four to one.

That single coastal representative wasn't bound to support commercial fishermen, nor were inland commissioners bound to oppose them. But experience in other states—such as Texas and Louisiana—had shown that when such statewide regulatory panels were dominated by sportsmen, as they usually were, they couldn't help but allocate the resource to sportsmen.

So, as some coastal states recognized the value of their marine resources, they authorized more specialized bodies to manage them. Unlike the "bubba boards," whose upcountry members could be insensitive to the realities of life on the coast, the new panels were comprised completely of saltwater people.

The Florida Marine Fisheries Commission provided a blueprint. Established in 1983 by Florida's legislature, the designation of the commission's membership was itself patterned on that of the regional management council system that Congress had established in 1976 for the management of federal fisheries: It was to be balanced among seven "knowledgeable" people and initially included two commercial fishermen, two sportsmen, two scientists, and one environmentalist.

In 1994, Mississippi's Legislature authorized a similarly structured Mississippi Coastal Commission after commercial fishermen complained that the statewide Commission on Wildlife, Fisheries and Parks was not addressing their needs.

"At Parks & Wildlife, they had the 'sport' mentality, you know, of deer and turkeys, bass and all that stuff," said Jean Williams, an industry advocate. "And every time we wanted something regulated, we had to go all the way up to the Capitol in Jackson—about 200 miles away— and they didn't understand what was goin' on. So, we put in to get this 'seafood commission' to come back down on the coast, and the Legislature approved it."

(A similar "Mississippi Marine Conservation Commission" had in fact been established by the legislature in 1960, after it abolished the former Seafood Commission. The Marine Conservation Commission had a biologist who was also a member of the ten-man panel; he carried six votes on biological matters and could not often be overruled. That body was later abolished as well.)

Williams was married to a commercial fisherman and had been a prime mover behind the founding of the Pascagoula-based Save America's Seafood Industry (SASI), the local group that would rally the fishermen's defense against the CCA's coming blitz.

SASI had been instrumental in determining the makeup of the new coastal commission. According to Williams, it was "made up totally of coastal people from the three coastal counties, and each county has two people. Jackson County, where I live, the 'seafood processor' came from, and the non-seafood 'at-large industry' person came from. That's Doug Horn, who owns Clark Seafood in Pascagoula, and William Mitchell, who owns a car dealership, William Mitchell Motors, in Pascagoula.

"Over in Harrison Country, around Biloxi and Gulfport, we have the 'charter-boat industry' rep, Henry Boardman. And Sherman Muths is the 'recreational fishing' person. He's an attorney, a millionaire from Gulfport, and the chairman of the commission. Then Hancock County has the representation of the 'commercial harvesting' industry and the 'nonprofit environmental' group. There we have Oliver Sahuque, a shrimper from Lakeshore, and Dr. Vernon Asper of Diamondhead, an oceanographer and Sierra Club member."

In addition to those appointees, the seventh member of the panel was the coastal Fifth Congressional District's sitting representative on the statewide Commission on Wildlife, Fisheries and Parks. Recreational fisherman Webb Lee of Woolmarket, a Biloxi suburb, filled that position.

In a statement following his appointments, Governor Kirk Fordice said, "The law creating this commission requires that the appointees represent six groups that have a vested interest in the coast's aquatic life. It is my firm belief that all Mississippians have a vested interest in our precious and treasured coastal waters, and I firmly believe that my appointees will represent our entire state well in that regard."

The new seven-member Mississippi Commission on Marine Resources had regulatory authority over all coastal aquatic issues including recreational and commercial fishing, wetlands protection, and coastal zoning. With its balanced membership and emphasis on science, the CMR set out to demonstrate that it would be responsive both to the public and the needs of the marine environment.

After appointing an interim director and initiating a nationwide search for a "true professional" to lead the new Department of Marine Resources, which housed the state's coastal scientists, the Commission

DEATH BY A THOUSAND HEARINGS

Red drum: The most controversial fish in the Gulf.
(Painting by Kathleen Joffrion)

on Marine Resources took over on July 1, 1994.

With the Florida net-ban referendum still four months away, Mississippi sportsmen weren't yet clamoring for a ban of their own, so the panel wasn't immediately faced with the Solomonic task of resolving that issue. Still, the coastal commissioners' first foray into fishery management concerned the most controversial species of fish in the Gulf—red drum.

RED ALERT!

When New Orleans chef Paul Prudhomme popularized blackened redfish in the mid-1980s, a few purse seiners targeted the older fish that schooled offshore. As biologists sampled the seiners' catch, they learned that the population of breeders, which had never been harvested in significant quantities, had nonetheless been declining for years.

Managers in the Gulf states had allowed their sport and commercial fishermen to catch too many of the smaller redfish that spent the first few years of their lives near shore.

The purse seine fishery occurred primarily in offshore waters, which opened the door for federal regulation. To address the overfishing and rebuild the offshore population, federal managers advised the states to tighten their restrictions to allow 30 percent of each annual crop of juvenile redfish to escape the inshore waters.

In 1990, Mississippi's managers increased the minimum recreational size limit from 16 to 22 inches. Four years later, the inside waters appeared to be teeming with young redfish. Hundreds of anglers, who wanted to bring some of them home, signed petitions that called for returning the minimum to 16 inches.

In May 1994, the Commission on Wildlife, Fisheries and Parks, which was still managing the fishery, acceded to their wishes.

Commissioner Webb Lee, who would later represent the commission on the new coastal panel, led the charge for the size reduction.

State fishery biologists, however, strongly opposed the reduction: "Lowering the limit to 16 inches will not allow the necessary escapement rate," stated Tom Van Devender, chief of saltwater fisheries for the state Bureau of Marine Resources. (The BMR housed the DWFP fishery scientists. When the new coastal commission took over, this division was relocated to Biloxi and renamed the Department of Marine Resources.)

"There will be an increase in mortality that could jeopardize the whole program," said Van Devender, at the commission's May meeting in Jackson. "Sure, there would be a good catch for a couple of years until we depleted the current crop of fish, but after that, all you'd catch would be the small fish just like you did before the regulations."

James "Tut" Warren, of the University of Southern Mississippi's Gulf Coast Research Laboratory, in Ocean Springs, added that it was the 22-inch minimum size limit that had enabled the escapement rate to reach—and considerably exceed—the federally mandated 30 percent. A reduction to 18, he said, would allow recreational anglers to catch more fish while still maintaining an acceptable escapement rate at or slightly above the 30 percent mark. A reduction to Lee's politically popular 16 inches, however, would be disastrous.

"In periods of low abundance [due to environmental factors], you will increase the nearshore mortality rates," Warren warned the commissioners. "The mean length of fish that were being caught prior to 1990 was about 16 inches. If we revert back to 16 inches, then we will see basically the same number of fish being caught as there were prior to 1990. When that was happening we saw that we had a very low escapement rate. During some periods, over 90 percent of these smaller fish were being caught. That left very few fish to survive long enough to make it offshore to maintain breeding stocks."

Commissioner Lee, a former elected Clerk of the Circuit and County Courts in Harrison County, dismissed the scientists' evidence: "Fishermen know more about what's out there than biologists do. If anyone believes fishermen are happy with 22 inches ... they haven't talked to fishermen."

Ray Lenaz, the chairman of Mississippi's CCA chapter, considered the proposed size reduction "great if we can reduce the commercial catch at the same time."

Lenaz, a Biloxi sport fisherman, had originally opposed the 16-inch proposal but said he thought there were enough sport fishermen practicing catch-and-release to offset the decrease in the size limit: "Just because there is a three-fish [daily bag] limit, it doesn't mean everyone has to keep three fish every time they go out." Lenaz did allow that, if the biologists' monitoring indicated that the fish's recovery was being jeopardized by the smaller size limit, "then I'll be the first to say we need to change back to 18 or even 22 inches."

Another active CCA officer, Gulfport angler Pete Umbdenstock, didn't object to the 16-inch minimum, "as long as they don't change the creel limit from three."

(In states with more sizeable fisheries, like Florida and Louisiana, the Coastal Conservation Association's chapters generated enough funds to hire full-time executive directors. In Mississippi, the state chapter was led by a revolving tag team of volunteer officers.)

Commissioner Lee's four colleagues on the statewide panel initially voted against his proposed six-inch size reduction and approved, instead, the more conservative 18-inch minimum. Still, in the final moments of the meeting, Lee brought his proposal up again, this time with a sweetener: To make up for the increase in recreational harvest effected by a reduction to 16 inches, take some fish from the commercial fishermen by limiting their daily take to just three reds, the same number that sport fishermen were allowed.

Lee's fellow commissioners found his late-breaking proposal more palatable and, turning their backs on their own experts, rescinded the 18-inch limit that they'd just passed, shaved off the two inches Lee wanted, and added his proposed restriction on commercial fishermen.

The about-face stunned the commercials. "Mr. Webb Lee, I can't understand where you're coming from," said SASI's Jean Williams. "You tell me how a man can make a living the same way a man can go play?"

Mississippi's commercial fishermen were the last on the Gulf Coast still able to harvest wild redfish for the market.

Texas, where the Coastal Conservation Association got its start, had made the species a gamefish in 1981; Alabama followed in 1984, then Florida and Louisiana in 1988.

During the 1990 cutbacks, Mississippi had initiated a season-and-quota system for the commercial sector: The season opened on October 1 and fishermen could land as many reds as they were able until an industry-wide quota of 35,000 pounds was reached, usually by late winter.

By comparison, the recreational fishery was conducted year-round with no hard quota; the catch was controlled by size and daily bag limits on individual anglers. Under that regime, sportsmen in 1994 landed 300,000 pounds of red drum and unintentionally killed an unspecified number of "regulatory discards," which had failed to meet the size requirements and died after they were returned to the water.

Commercial fishermen, who were responsible for about 12 percent of the total annual harvest, considered their quota a pittance and wanted it doubled. Restricting them instead to only three reds per day, Williams told the commissioners, was "absolutely ridiculous."

Prior to becoming law, Lee's redfish proposal had to be aired in a round of public hearings and then put to a final vote by the commission.

At the June 20 hearing in Biloxi's J.L. Scott Marine Education Center, Commissioner Lee reiterated, "I'm obligated to the little man fishing off the pier or seawall. I am committed to the 16-inch redfish." In his support, a lady asked the commissioners, "I've got children and grand-children and how can you tell a child or grandchild who catches a 16-inch redfish that you have to throw it back?"

Sportsmen expressed mixed opinions, with many still favoring the reduction to 16 inches while others supported the more conservative reduction to 18 inches.

Commercial fishermen stayed out of the size-limit debate but roundly opposed the proposed daily limit on themselves. "You just can't make enough money on three redfish a day," said one, while Pascagoula netter Hilton Floyd told commissioners, "This is about driving another nail in the coffin of the fishermen."

"There's a happy meeting ground somewhere here," said Jim Walker, public relations director of the state Department of Wildlife, Fisheries and Parks, which still housed the division that enforced fishery laws on the coast. Walker voiced his agency's support for the proposed cut in the commercial limit, presumably on the basis that its field agents couldn't distinguish sport from commercial fishermen: "The ease of enforcement is one reason to require just three redfish caught per day for both recreational and commercial fishermen."

On June 30, the day before the new coastal commission was to assume control of marine fisheries, the five-member Commission on Wildlife, Fisheries and Parks met in Jackson and in one of its last acts, voted to establish the popular 16-inch minimum without any additional limits on the commercial fishery.

The regulation wouldn't stay in place long enough to hurt the redfish population.

When the new Mississippi Commission on Marine Resources took over management of the state's marine fisheries, its members were of course mindful of the adverse impacts that inadequate size limits in the recreational fishery had had on the red drum population and didn't want a repeat under their watch. At the panel's second monthly meeting, in August, the commissioners approved a notice of intent to raise the minimum size limit to 18 inches and allow sportsmen to retain one "bull red" per day, over 30 inches in length.

The measures would allow anglers to bring in some more redfish, yet not so many that the recommended escapement target would be exceeded. The public generally voiced approval of the commission's balanced proposal during September hearings and the panel finalized the redfish rules at its October meeting.

The following month, Floridians voted to ban nets. Mississippi's sportfishermen clamored for the coastal commission to do the same.

FLORIDA FALLS

On November 8, 1994, Florida's voters overwhelmingly approved the recreational industry's ballot measure that eliminated most finfish nets and shrimp trawls from the state's nearshore waters.

In a November 20 article, "avid sportfisherman" Pete Umbdenstock told the Biloxi *Sun Herald* that the lopsided 7:3 vote totals in Florida sent a message that Mississippi's legislators needed to heed. "The Florida Legislature wasn't being responsive to the will of the people, and it is a sad state of affairs when that happens," he said.

"And, quite frankly, I don't think the Mississippi Legislature—and most especially some of our coastal delegation—is being responsive to what the majority of its constituents want either."

GCCA's state chairman Ray Lenaz agreed with his fellow officer's claim that most Mississippians wanted to do away with netting. "There's no doubt that if we had a referendum on this subject it would pass. Some of our legislators are just not listening," he said.

Given the sportsmen's numbers, and pop media's jaundiced view of commercial fishing, it was possible that a majority of Magnolia State voters would have gone along with a ban on netting fish down on the coast. Still, this was Mississippi, not Fantasyland.

More than one-third of the state's 2.6 million residents were Black and plenty of those folks had been supplementing their diets with offerings from the wild for generations. If they connected the dots between their fish fries of crispy—and inexpensive—mullet, croakers, and spot with the nets that provided them, who knew?

Mississippians, however, would not be voting on nets, despite multiple threats by the sportsmen to put the issue on the ballot.

Resolution of the netting question fell to the coastal commissioners, who were knowledgeable, experienced, and expected to base their decisions on the best available science. CCA leaders, who lacked any veritable scientific justification for banning nets, fanned the flames of hysteria.

If the nets weren't banned immediately, the fish would go extinct, they repeated, a claim that jibed with the rhetoric that had been raining down from the national nonprofits and their funders. Another of the anglers' justifications for a ban was harder to rebut because there was at least some truth to it—the state could be invaded by some of the netters that Florida's voters had just put out of business.

What would a working fisherman do if he was forced to hang up his nets? Probably break into another fishery, like crabbing, and remain at home where he had roots and was familiar with the waters. If he was locked into full-time net fishing, little Mississippi would hardly be a preferred destination compared to, say, North Carolina or south of the border.

Still, it was an unknown, and CCA was hyping the threat in Mississippi and in the other central-Gulf states.

In fact, the Mississippi Legislature had already acted on the matter: In a special session earlier in the year, it passed a law that basically said that if a resident of Mississippi couldn't do something in one state (net fish in Florida) then residents of that state (Florida) couldn't buy licenses to do it (net fish) in Mississippi.

Because it interfered with interstate commerce and targeted a single class, that law might not withstand a court challenge, suggested GCCA's

Lenaz. If it did not, he told the *Herald*, the only way to "effectively thwart a surge of Florida netters" and protect the state's "already beleaguered fish stocks" was an "all-out net ban."

Such a ban would not only rectify the "long-term effect of overfishing" that was, in his opinion, caused by netting, it would have an immediate positive effect for the "pier fisherman, the wade fisherman, the guy standing along the rocks at the harbors, and the small-boat fisherman" who couldn't get away from where the nets were "wiping out everything around them."

It was not true that net-ban advocates were trying to destroy the livelihoods of a few people, "as we've been accused," said Lenaz. "We are talking about protecting the future for everyone, the hundreds of thousands who, every year, enjoy Mississippi's natural resources and who contribute far more to the economy than commercial net fishing ever dreamed of doing."

There were some strong figures to support his stand, added John Lambeth, the author of the *Sun Herald* article. In 1994, the state issued 132 gill-net licenses while the U.S. Fish and Wildlife Service estimated that more than 100,000 people were fishing recreationally in Mississippi each year. "Economically, state compiled figures show gill-net fishing accounts for less than $500,000 in economic impact versus $125 million to $150 million for sportfishing on the Coast."

According to CCA's Lenaz, "A ban right now would do the least amount of harm to a very few and most amount of good for an awful large majority of users."

One of those "very few," Hancock County commercial fisherman Don Joost, told the newspaper that the issue boiled down to allocation. "It's one user group trying to take away from another," he said. Regulations should be based on scientific data and no group should be excluded from a fishery, added Joost.

"I got just as much right to that water out there as any sportfisherman as long as I do it within the regulations."

The commissioners postponed a vote on the net-ban issue at their November meeting until they could learn more about it. As part of their education, they held a couple public hearings in Biloxi where they got an earful.

SEAFOOD CAPITAL OF THE WORLD

Biloxi had a population of about 50,000 and was the state's third largest city, after Jackson and Gulfport.

The coastal city had already been under the rule of the French, English, and Spanish when Mississippi joined the Union in 1817 as the twentieth state. A melting pot from the beginning, Biloxi attracted successive waves of international immigrants with one common language—seafood.

By the 1820s, Dalmatian fishermen from the Adriatic coast of what is now Croatia, who had crossed the Atlantic to escape Austrian rule, had worked their way down from East Coast ports to Mississippi. The Dalmatians adapted their square-sailed luggers to the shoal waters of Mississippi Sound by adding retractable centerboards, and the low-sided, wide-beamed vessels made ideal platforms for tonging oysters. The Slavs worked the Sound's many natural reefs, and later pioneered the Gulf Coast's cultivated oyster fishery.

Meanwhile, Mediterranean immigrants from France, Greece, and Italy had begun to seine shrimp around Biloxi. They first used shallow-drafted catboats but as the industry grew so did the size of their boats, which led to the design of the Biloxi schooner. Shallow-draft and beamy, these grand vessels ranged to more than 70 feet in length yet could navigate in a few feet of water. And with the two masts rigged with up to six sails, they could swiftly transport fresh product back to port before it spoiled.

These Biloxi schooners, moored at the dock of the C.B. Foster Packing
Company, in the late 1930s, were rigged for oyster dredging.
(Tony Ragusin photo courtesy of Val Husley, Maritime and Seafood Industry Museum)

In the early 1870s, a new railroad between New Orleans and Mobile opened new markets, while a technological advance in food preservation catalyzed Biloxi's boom in seafood processing.

After the Civil War, George Dunbar, a shoe salesman from Massachusetts who'd relocated to New Orleans, began to adapt French fruit-canning techniques to seafood. Early results were disappointing—natural acids in the shrimp reacted with the metal to turn the crustaceans black and corrode holes in the cans. By 1880, Dunbar had learned to line the containers with parchment and his George W. Dunbar & Sons company was turning out an appealing product. The following year, a group of five entrepreneurs opened the first seafood cannery on Mississippi's Gulf Coast.

Biloxi is situated on a peninsula, with the Mississippi Sound in front and the protected Biloxi Bay in back. Lopez, Elmer and Company built its plant on the back bay where it canned oysters and shrimp, handled raw oysters in bulk and, during the off-season, preserved figs. Encouraged by that firm's success, a group of investors from New Orleans and Biloxi organized the Barataria Canning Company in 1882 and built a plant on Point Cadet, at the tip of the Biloxi Peninsula. In 1884 came the Biloxi Canning Company; Lopez, Dunbar's Sons and Company; and the E.C. Joullian Packing Company. In 1886, William Gorenflo and Company opened its doors.

The population of Biloxi more than doubled from 1,500 in 1881, when the processing industry started, to 3,234 in 1890. Still, the demand for labor outstripped the supply.

The Sea Coast Packing Company imported the first group of transient workers from Baltimore, where Chesapeake Bay's immense crab and oyster crops were processed. These immigrants from Central Europe—mostly Poles and Yugoslavs along with some Czechs, Austrians, and Germans—were known locally as "Bohemians."

Vintage label on shrimp canned by G.W. Dunbar's Sons, a branch of Dunbar's, Lopez & Dukate Co. of New Orleans and Biloxi.

Before long, the other plants were importing their own workers from Maryland. Brought in by train at the company's expense and housed at labor camps that were built close to the factories, the men did the oyster shucking, dock work and fishing while the women and children peeled shrimp in the factories. When fishing season ended, some of the Bohemians returned to Baltimore while others chose to remain in Biloxi.

By 1892, oyster and shrimp canning had become Biloxi's chief industry. Orders for canned products poured in from across America and the capitals of Europe, and the city's factories turned them out by the trainload.

Lopez, Dunbar's Sons and Company was the most successful, with over 600 workers during the peak season. The company also had extensive holdings of private oyster beds in the Sound, a fleet of 60 large steamers and schooners as well as hundreds of two-man sailing boats that were used to tong oysters in Louisiana's marshes.

The enterprise evolved into the Lopez and DuKate Company, which in 1901 shipped out over 525 boxcar-loads of canned shrimp and oysters.

In 1902, according to University of Southern Mississippi history professor Deanne Stephens, the city's 12 canneries processed a catch of nearly 6 million pounds of oysters and more than 4.4 million pounds of shrimp.

In 1903, Biloxi claimed the title of "The Seafood Capital of the World" when it eclipsed Baltimore as America's leader in seafood canning.

Biloxi's oyster production in 1905 was tops in the nation at $1.34 million (about $48 million in 2025). Maryland produced nearly $550,000 worth of oysters that year, and Louisiana just over $500,000.

At least 1,600 of the city's 10,000 residents were employed by the seafood industry in 1910 when the *Biloxi Daily Herald* reported: "It was very much in evidence on the streets of Biloxi Saturday night and especially on Howard Avenue that a wave of prosperity had struck the city, as the streets were packed and jammed as they were never before with people sight-seeing and buying the necessities of life. Everybody seemed to have a good supply of ready cash and the merchants report good business."

In 1912, an unusual vessel that would change the course of the seafood industry chugged into Biloxi Harbor—W.P. Kennedy's Bernadino was powered by a gasoline engine and rigged with an otter trawl.

Factory-owned shrimp seines, which measured up to 1,200 feet long and 18 feet deep, required several skiffs and a large crew to handle, yet a 30- to 35-foot trawl could simply be towed behind a motorized vessel and worked by just two men.

Fishermen had begun to organize and strike for higher wages as early as 1903, and the new harvesting technology accelerated the shift away from the old system of fishermen working as factory employees, and toward the modern system of independent harvesters marketing their catch to the highest bidder.

Oyster and shrimp boats in Biloxi's Back Bay in the 1940s. The third vessel from the right is a schooner. Most of the wider boats are "luggers" that were originally rigged with single lug sails, then converted to power. Most of the tall uprights appear to be masts for sails but are more likely booms with blocks that were used to pick up the shrimp trawl so the catch could be dropped on deck and sorted.
(Anthony Ragusin photo courtesy of Mississippi Department of Archives & History)

Just prior to World War I, Biloxi's canneries were turning out 15 million cans of oysters and shrimp a year. The war in Europe crippled the U.S. economy and suppressed demand for specialty items like canned shrimp and oysters. At the same time, the industry work force declined as the U.S. government restricted Slovenian immigration, which essentially halted in 1917. To fill the void, the canneries brought in French-speaking Cajuns from Louisiana. The first group of Cajuns that was recruited en masse arrived in Biloxi in 1914; over the next 15 years more moved to Biloxi and other Mississippi coastal towns after Louisiana's sugar cane crops were devastated by disease.

After World War I, the seasonal migration of workers from Baltimore ended and the days of the vast canning empires were numbered—of the original big five, only the Biloxi Canning Company remained in 1934, and it eventually went under as well.

The golden age of the big packing plants was over, yet Biloxi's processing industry continued to develop: Instead of the sprawling canneries with their factory-owned fleets and imported labor force, latter-day companies purchased their raw materials off the boats of independent fishermen or from other dockside buyers across the Gulf Coast. Modern factories used the latest automation technology to process those shrimp and oysters, crabs, and fish into the fresh, frozen, or breaded products that consumers had come to prefer.

The amount of seafood trucked into Mississippi for processing far exceeded the state's harvest, and nearly all the finished product was exported from the state. While the total retail value of the processed seafood was many times that of the raw materials, the processors' whole economic impact extended beyond the value added to the fishermen's harvest to include expenditures on packaging and other supplies, wages, utilities, transportation, and taxes.

In the 1970s, another wave of immigrants gave the industry a shot in the arm. After South Vietnam fell to the communists in 1975 many of its refugees, who'd traditionally fished in their country's coastal waters, were directed to fishing communities on the Gulf Coast.

The first small group to arrive in Biloxi took jobs in the packing plants. The refugees pooled their money and by the 1980s had begun building and operating their own vessels and opening their own markets and processing plants. By 1983, the city's Vietnamese population numbered roughly 2,000, and eventually most of the participants in Biloxi's seafood industry would be Vietnamese.

DEATH BY A THOUSAND HEARINGS

In 1986, nearly 700 workers were employed in Biloxi's 17 seafood processing firms, a cat-food processing plant, four boatbuilding and repair yards, two marine fabrication and construction businesses, and three net and trawl shops. That the city's leaders continued to value productive enterprise was evident in their 1985 Waterfront Master Plan, which identified several locations with "good highway, water, and rail access," that were suited for "seafood processing and other marine-oriented industry."

Then, in June 1990, the Mississippi Legislature legalized gambling.

FROM CANNERY ROW TO CASINO ROW

At the same time the hardworking men, women, and children drudged on the docks and in the steaming seafood factories along Biloxi's Back Bay, a more genteel set amused themselves elsewhere on the same peninsula.

Wealthy New Orleanians and planters from the state's interior summered on the breezy coast in Gulf-front hotels and their own oak-shaded antebellum mansions.

The Civil War had set back the region's development as a tourism Mecca but, by the 1870s, visitors from across the South were arriving via the same new railroad tracks that carried the seafood companies' products to distant markets.

The freewheeling lifestyle on Mississippi's coast was a world apart from the austere Bible Belt culture of the interior, where drinking and gambling were frowned upon and laws against the sinful practices were generally respected.

Mississippi passed its first statewide Prohibition law in 1907 and in 1919 it was the first state to ratify the federal Volstead Act, which prohibited alcohol sales nationwide.

The author of the bill that ratified Prohibition said that the victory over alcohol had nowhere "been more marked and complete than in Mississippi, which, through a brave, honest, law-loving, home-loving legislature, drove the legalized traffic from the whole state."

Prohibition may have driven the *legal* sale of booze from the state, but it enriched the smugglers, bootleggers, and owners of speakeasies and resorts that catered to that segment of the population that still liked to wet their whistle.

In 1926, three Biloxi partners opened a luxury resort on Dog Key, an eroding barrier island that they renamed the Isle of Caprice. Alcohol could be served legally on the island because it was in territorial waters, beyond the state's jurisdiction.

Excursion boats ferried patrons from Biloxi to the offshore resort where cocktails were always on the menu at the restaurant and a casino offered roulette, craps and other games that were forbidden on the mainland. People flocked to the "Monte Carlo of the South" from across the country. Later in the 1920s, engineers stabilized the peninsula's frontage on the Gulf with a concrete sea wall; to buffer it from storms, they pumped enough sand from offshore to create a gently sloping beach that would eventually stretch nearly 30 miles to become—according to the Biloxi Bay Area Chamber of Commerce—the world's longest man-made beach. With its lazy surf and glistening white sand Biloxi Beach became a tourist magnet.

In 1925 the road known as the Old Spanish Trail was modernized into U.S. Highway 90. The coastal east/west route would eventually be expanded to four lanes and extend from Jacksonville, Florida, to west Texas.

Route 90 hugged the beach around Biloxi and nearby Gulfport where the scenic stretch of "Beach Boulevard" sprouted restaurants and clubs that, despite Prohibition, somehow managed to keep well stocked with booze.

DEATH BY A THOUSAND HEARINGS

In 1929, the motorized vessel Mercedes Williams was found stranded on a reef in Galveston Bay with $75,000 worth of liquor in her hold. The vessel was built in Biloxi and licensed in the name of the DeJean Packing Company.

In 1932, the Coast Guard seized the Freda, a motorized schooner, off Pass Christian. Owned by the Biloxi Canning Company the boat had a cargo of contraband liquor valued at more than $60,000. The Freda had allegedly been on an oystering trip to the Pass Christian reefs.

Prohibition became increasingly unpopular during the Great Depression. The repeal movement, led by conservative Democrats and Catholics, emphasized that legalizing the sale of alcohol would generate enormous sums of much needed tax revenue and weaken the base of organized crime.

Congress repealed the federal act in 1933 and allowed the states to set their own laws for the control of alcohol. (Mississippi would be the last state in the nation to repeal Prohibition, in 1966.)

In 1938, the legislature in Jackson outlawed slot machines and pinball machines. Down on the coast, the state's ban on gambling proved about as effective as its ban on the sale of alcohol.

In 1941, the owner of Biloxi's Art Deco Broadwater Beach Hotel was fined $75 for possession of a dice table, roulette wheel and faro layout, and $150 for having six slot machines.

In 1951, two Biloxi ministers traveled to New Orleans to testify before the federal Kefauver Crime Commission, a panel of U.S. senators investigating organized crime. They told the panel that slot machines were so pervasive in Biloxi that there was one per every 35 residents of the city. Blackjack, dice, and other gambling activities were flourishing as well, they said.

The ministers also pointed out that the average age of soldiers at the nearby Keesler Field was nineteen, an age at which they had not matured to avoid "places of iniquity."

After World War II, the Mississippi coast became a popular vacation destination. Hotels, restaurants, and night clubs sprang up along Route 90 through the 40s, 50s, and 60s. Big-name entertainers on "The Strip" included Hank Williams, Jane Mansfield, Jerry Lee Lewis, Fats Domino, Elvis Presley, and popular southern comedian Brother Dave Gardner.

B-drinking—the employment of female workers to solicit drinks from male tourists—was widespread in bars and strip joints, as was gambling, prostitution, and other vice, all conducted under the "protection" of the local police. After 1969's Hurricane Camille flattened nearly everything along the Mississippi coast, it was rebuilt as a more wholesome destination for families and conventions. But folks who enjoyed gambling could still find some action.

In 1987, the Europa Star was moored at Biloxi's Point Cadet Marina. The 167-foot cruise ship operated in federal waters where it featured dining, entertainment, and gambling.

In 1988, Noel J. "Jimmy" Skrmetta began to run gambling cruises to federal waters from his Pier Seafood Restaurant on Biloxi's waterfront.

After the U.S. Congress allowed recognized tribes to open casinos, in 1988, the gambling industry began to push for more widespread legalization.

Conservative residents of southern states wanted the practice to remain confined to the reservations, to protect the "quality of life" and "moral values" of their own communities. Alabama strictly prohibited most forms of gambling in its constitution.

Early in the 1990s, politicians in both Louisiana and Mississippi found a way to introduce gambling without alienating too many of their constituents by restricting it to facilities on or near water.

DEATH BY A THOUSAND HEARINGS

Casinos and beach on Biloxi Bay in July 2021. When gambling was legalized, it was initially restricted to floating vessels but later expanded to onshore casinos. *(Michael Warren photo from iStock/Getty Images)*

Mississippi's 1990 Gambling Control Act gave voters in the counties along the Mississippi River and the Gulf Coast the freedom to legalize gambling. By 1992, two of the state's three coastal counties—Hancock and Harrison—had approved dockside gambling.

Legalizing gambling in Harrison County ignited the biggest economic boom in Biloxi's history.

By 1994, Biloxi's Gulf and Back Bay waterfronts had 11 floating and onshore casinos, which would eventually consolidate into at least seven major "destination" casino resorts.

Property values soared as the casinos acquired and developed the dockside locations that the law stipulated. The uniquely water-dependent businesses that were traditionally associated with the seafood industry—

such as dockside buyers, processors, ice houses, boatbuilding, and repair facilities—were rapidly displaced.

Sea Grant extension agent David Burrage, of Biloxi, bemoaned the loss: "To me it is ironic that the fishing fleet is water dependent. The things that have taken their place, the casinos, are not water dependent. They're only on the water because of the law." Burrage thought more should be done to help the seafood industry. "People don't realize the lure, the aesthetic appeal of a working waterfront," he said.

"At the Grand Casino in Gulfport, people love walking down to the boats to buy shrimp right off the dock. It's neat to come into a place and see a shrimp boat. There are places that would pay to have those types of attractions."

Leaders of the Coastal Conservation Association's Mississippi chapter held an opposite view: Since Biloxi's seafood-industry infrastructure was already being squeezed out, maybe it was time to just go ahead and eliminate not just the gillnets but *all* commercial fishing from the Sound.

With the seafood industry out of the way the coast could become a tourism utopia, or so claimed a late-1994 newsletter of the conservation group: "We could designate our coastal waters inside the barrier islands for sport fishing only and promote the area nationwide. Sport fishing, casino gambling, golf, clean beaches, and our excellent restaurants would make a 'hard to beat' entertainment package for our tourism commission to promote. Visitors would actually be able to catch fish from the beach in front of their motels. ... Money and jobs lost in the seafood industry would be more than made up for in the tourist and sport fishing we could all enjoy. ...We might even use the gambling revenue and/or some of the money generated by the saltwater fishing license to retrain and outfit people who make their living from commercial fishing to operate sport fishing charter boats. We would sure need them with the improved sport fishing and increase in tourism."

COMMISSION ON MARINE RESOURCES HEARING IN BILOXI

Biloxi was the logical location for the Commission on Marine Resources to hold its public hearings: The city was centrally located on the coast, had the facilities to hold a large crowd and was the new home of the state's marine resources agency.

The commission's December 7, 1994, hearing was the first of two that the panel planned to hold on the net-ban issue; the proposal under consideration at this one didn't call for a *statewide* ban—it would eliminate gillnets and trammel nets from the waters off Harrison County only.

Even so, 750 people turned out. Biloxi's J.L Scott Marine Education Center seated about 300, so most of the crowd milled outside.

Inside, Tom Devender, a fishery biologist with the Department of Marine Resources, used charts and an overhead projector to show the attendees the status of the state's three most contested species.

Devender's presentation didn't jibe with the hysterical claims of local sport fishermen or the "overfishing everywhere" propaganda campaign of the philanthropies and national environmental and media organizations. He told the crowd that the populations of the state's redfish, trout, and mullet were healthy, the species were being managed sustainably, and there was no biological reason for banning the commercial fishermen's nets.

Like water off a duck's back, Devender's presentation had no effect on the Coastal Conservation Association's members and the political entities the group had captured.

The *Sun Herald*, in a late-November article captioned, "Florida net ban sending waves along Gulf Coast," noted that a certified public accountant from Gulfport, "with support from the Gulf Coast Conservation Association," had "in recent weeks ... championed a drive to enlist the support of local governments to halt gill-netting in Mississippi waters."

Most of those "local governments" took the CPA's bait: Biloxi's City Council sent a representative to the hearing to ask the commission to enact a ban; so did that of sister city Gulfport and the city councils of Pass Christian, Long Beach, Bay St. Louis, and Ocean Springs.

Harrison County's Board of Supervisors also sent a representative to the meeting; he told the panel that a net ban would "promote conservation of the natural resources of Harrison County" because "the use of purse seines, gillnets, trammel nets, and other entanglement nets is becoming more common along the coastline of Harrison County; and this type of fishing entraps all kinds of marine life and threatens to deplete and possibly destroy fishing along the coastline in Harrison County and the adjoining counties of Jackson and Hancock; ... ."

(Despite these ominous predictions, neither of the "adjoining counties" would follow Harrison's lead: Hancock County's supervisors had already tabled GCCA's resolution to outlaw net fishing, and so would those in Jackson County.)

The Mississippi Department of Wildlife, Fisheries and Parks, which was of course directed by a commission with no commercial representation, joined the GCCA's charge against netting. The agency's public relations director told the panel, "The Mississippi DWF&P supports any efforts the Department of Marine Resources can take to enact a ban on the use of gillnets, trammel nets or any kind of entanglement devices within state marine waters."

Possibly smarting from the recent transferal of its management authority to the more focused coastal commission, the Jackson-based agency demonstrated why commercial fishermen were so keen to get away from it in the first place: In addition to its presentation at the hearing, the department issued an official press release that argued for the sportsmen's proposed ban and was reproduced verbatim in several newspapers in Mississippi, as well as in neighboring Louisiana, where GCCA was rallying sportsmen for a springtime assault on the Bayou State's legislature.

The following example appeared in the December 8 *Ocean Springs Record:*

"The DWF&P is joining a growing group of Coast cities who have passed resolutions supporting a proposed ban. Bay Saint Louis, Long Beach, Gulfport, Biloxi and the Harrison County Board of Supervisors have all gone on record in support of the ban.

"Texas, Georgia, South Carolina and California have banned the use of these nets in their waters. In November, Florida—by a wide majority of the popular vote—also banned these nets.

"The DWF&P is concerned about both the impact these nets have on Mississippi's marine resources and the difficulties associated with enforcing the existing regulations.

"The DWF&P Marine Enforcement Division is responsible for enforcement of all DMR regulations in Coastal waters. The Marine Division has arrested seven parties for gillnet violations in the past 14 days. All seven violators were using gillnets to catch roe mullet. Six were arrested for netting in closed waters.

"Enforcement of the present regulation is very difficult. The law states that DWF&P officers must catch violators with their nets in the water. 'Strengthening the existing regulations is an option but we believe a complete ban would be the most effective means of solving the problem. Not only would the ban address the issue of resource protection, but it would eliminate the potential influx of netters from neighboring states to our waters,' DWF&P spokesman James Walker said.

"According to Terry Bakker, assistant chief of law

enforcement, the recent ban in Florida is already having an effect in Mississippi.

"'We've already seen an increase in out-of-state boats in our waters,' Bakker said. "The market for roe mullet is so lucrative and the penalties are so low that the illegal netters are willing to risk getting caught in closed waters and regard the fines as just the cost of doing business.'"

1994 COMPARATIVE LANDINGS IN POUNDS BETWEEN POPULAR **SPORT** AND **COMMERCIAL** FISHERIES

SPECIES	RECREATIONAL	COMMERCIAL
Red Drum	298,591 (88%)	40,246 (12%)
Spotted Seatrout	203,687 (74%)	73,034 (26%)
Sand Seatrout	323,278 (75%)	109,823 (25%)
Sheepshead	438,098 (66%)	228,910 (34%)
Flounder	141,647 (78%)	40,753 (22%)
Black Drum	141,615 (71%)	56,853 (29%)
Spanish Mackerel	61,140 (62%)	37,520 (38%)
Whiting ("Ground Mullet")	72,944 (48%)	78,614 (52%)
Striped Mullet	132,430 (14%)	781,300 (86%)
TOTALS:	1,813,430 (56%)	1,447,053 (44%)

In all, ninety-four speakers voiced their opinion on the proposed ban before public comments were cut off—48 were in favor of the ban and 46 were opposed.

Sea Grant's David Burrage, who had a decade of experience at the Mississippi State University Coastal Research and Extension Center, joined commercial fishermen in opposing the ban. In his testimony and accompanying handout, Burrage acknowledged that he was not a fisheries biologist, but that as a marine resource specialist he "could and did read reports such as stock assessments, species profiles, landings data, etc. I also obtain much anecdotal information regarding fisheries from my frequent contacts with commercial and recreational fishermen, various segments of the seafood industry and seafood consumers. I am also an avid sports fisherman."

Burrage reminded commissioners that they were charged with managing fisheries for the benefit of *all* Mississippians, "not just those with the ability to harvest them with a rod and reel or net." Since CCA's proposal would effectively eliminate the commercial fishery, "where would a non-fishing Mississippi resident buy fish to cook for dinner? Or go to a restaurant and order Mississippi fish?" he asked.

The proposed ordinance was more about allocation than it was conservation, explained the fishery agent, and it would allocate all the fish to recreational fishermen while letting "the commercial fishermen and seafood consumers in the state hang in the wind!"

If conservation were directing the proposal, "then it would make more sense to adjust recreational management measures, where the bulk of fishing mortality is generated," suggested Burrage, who supported his claim with a table that compared the recent landings of several popular species by the state's sport and commercial fishermen (*see opposite*).

FEDS PILE ON: PARK SERVICE SAYS NO TO NETS AND TRAWLS

The Gulf Islands National Seashore is a unit of the National Park Service and includes a pair of elongated islands off Florida's Pensacola Bay, four of Mississippi's five barrier islands—East Ship, West Ship, Horn and Petit Bois—and a mile-wide buffer around each of those islands.

On the day after the marine resources commission's December 7 hearing, Jerry Eubanks, superintendent of the Gulf Islands National Seashore, addressed a letter to Glade Woods, the newly appointed director of the Department of Marine Resources.

With the "recent increase in public concern over the use of gillnets," the Seashore wanted the state to know its stance on the issue, Eubanks wrote: "Due to the National Park Service's mandate to protect the natural resources, we feel we can no longer support the continuation of net fishing within the waters of the Seashore."

The "extremely fragile and easily impacted" seagrass communities around the islands were in decline, said Eubanks, with an attendant "loss of species diversity and abundance." Additionally, an "apparent reduction in available fish" had also been noted, "leading us to a heightened sense of concern over the marine resources."

With their grass beds, sandy flats and churning surf, the barrier islands were magnets for fish and fishermen, both sport and commercial. In the early 1970s, the state of Mississippi prohibited commercial netting around the islands during the summer, when sport fishing was at its peak, to accommodate recreational anglers. At the time, the state also prohibited shrimp trawling within half a mile of the mainland's shoreline.

Now the Feds were asking the state to extend its prohibition on near-shore shrimp trawling to all National Seashore waters, and to bar all finfish netting, permanently.

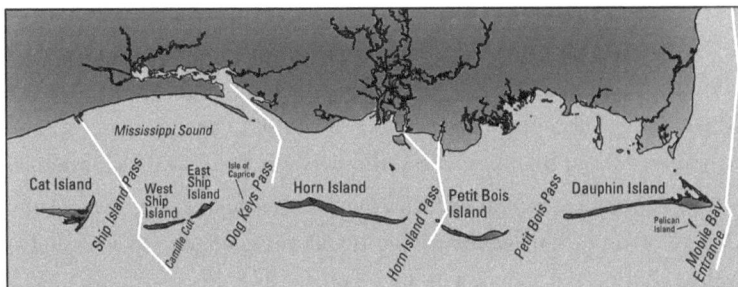

The Gulf Islands National Seashore encompasses all of Mississippi's barrier islands except the westernmost Cat Island. Dauphin Island, to the east, is an inhabited barrier island within Alabama's waters. The white lines mark channels that are maintained for deep-draft shipping.

"We do not suggest that a ban on net fishing and trawling will instantly return seagrass communities and fisheries populations to the levels experienced forty or fifty years ago," Eubanks acknowledged. "It seems apparent to all that the declines in the marine environment are far more complicated than that. We do feel however, that given the protection responsibility which the Seashore holds, we must develop a higher degree of protection than we are currently providing."

If the Park Service's proposals were based on a philosophical opposition to any commercial harvest within a national park boundary, that was one thing, said Mississippi's fisheries chief Fred Deegen, Ph.D. But as protective measures for submerged aquatic vegetation, they were uncalled for.

"I, personally, don't see that a one-mile distance is necessary," he said. "For one, the seagrasses simply don't occur on the south side of the barrier islands, in the open Gulf. They are limited to the north side of the barrier islands and certainly not such a great distance. An alternative to consider would be to close on the back side and leave open on the front."

The chief's suggestion would have kept seafood producers working in at least a portion of the Seashore's waters, which would in turn have allowed more members of the public to benefit from their resources. Maximizing public benefit was part of the state agency's charge, and managing for use by multiple stakeholders was one way to achieve that.

On *terra firma*, federal agencies like the National Forest Service and Bureau of Land Management typically permit a range of recreational as well as commercial activities on their lands. Multiple uses are sustainable although grazing, timbering, and other productive endeavors generally leave the lands in a less-than-pristine state than those within the parks. On the other hand, managing the parks for tourism alone leaves them ringed with development.

Net fishermen were hardly concerned with the differing management philosophies of sustainable use and preservation. They'd worked around the islands long before the National Park Service took them over in 1971, considered them to be essential fishing territory, and understandably resented the timing of the National Seashore's intervention.

In a sign of the times, nobody cared what they thought: The park service ended up prohibiting not just netting but *all* commercial activity—except for specially licensed charter-boat operations—within a mile of its islands. Fishing by private recreational anglers would remain unrestricted.

Cat Island, which was jointly owned by private parties and the state, was now the state's last barrier island open to commercial fishing.

CCA STRIKES OUT IN COMMERCIAL FISHING STRONGHOLD

After the coastal commission held its December 7 public hearing in Biloxi on GCCA's proposal to eliminate netting from Harrison County, the panel was expected to address the sport-fishing group's proposed *statewide* ban at its monthly meeting on December 19.

To turn up the heat on the commissioners, the anglers intended to show up with even more official supporters than they'd had at the first hearing. On December 12, they tried to add Jackson County's Board of Supervisors to their collection.

While tourism-happy Harrison County had sprouted casinos like mushrooms, the residents of Jackson County had never legalized gambling and forcefully opposed the opening of the county's single tribal casino in Ocean Springs.

On the Alabama line, in the state's southernmost corner, Jackson County is blue collar and industrial. The colossal Ingalls Shipbuilding— the largest employer in the state—was just one of several shipyards, and Chevron's refinery there was one of the biggest in the nation. The county was also a stronghold of commercial fishing.

Jackson County's fishing heritage was honored on the exterior of the 1940s county courthouse in Pascagoula with a striking bas-relief of three men hauling back a net.

Pascagoula, a town of 24,000, is situated near the mouth of its namesake river's main channel. The Pascagoula is a national treasure— it's the largest free-flowing river in the lower 48 states, unimpeded by dams, locks or levees. It's the only undammed river draining into the Gulf of Mexico from the United States.

With no impediments to their spawning migrations, rare Gulf sturgeon still abounded in the river.

The mouth of the Pascagoula River is divided into three branches, the East, Middle and West Rivers. They are connected by passes and bayous and merge into Pascagoula Bay, one of the state's largest and most productive estuaries.

From Pascagoula Bay east to Alabama's Grand Bay lies one of the most biologically productive estuarine ecosystems in the northern Gulf of

Mexico, which includes Mississippi's 18,400-acre Grand Bay Reserve.

Offshore of all this, the county's two barrier islands—Horn and Petit Bois—help stabilize salinity levels within the Sound and attract fish from nearshore and offshore. Over 90 percent of the state's marine fisheries harvest was landed in Pascagoula and adjoining Moss Point. In 1994, Pascagoula/Moss Point was ranked by the National Marine Fisheries Service as the nation's eighth largest port in volume of fishery landings (201 million pounds), and 34th in terms of value ($22 million).

Most of those landings were of the relatively low-valued (in terms of price per pound) menhaden, which are an indicator species for thriving wetlands and estuaries. Other products included shrimp, crabs, oysters, offshore reef fish such as red snapper, and of course, the inshore finfish netted by the small-scale operators that the anglers had come to ask Jackson County's supervisors to help them annihilate.

In mid-December, the fall mullet run was playing out, but a fellow could still get lucky and make a good lick if he didn't have to be at another damned meeting.

Forced to miss a night's work, fishermen were already not in the best of moods when Mississippi's CCA chief Ray Lenaz presented the supervisors with a bar graph of the state's historic commercial mullet landings. He pointed to a recent downturn and, implying that it was the result of overharvesting by commercial fishermen, asked the supervisors to pass his group's resolution to ban nets because, "With gillnets out there, we are going to overfish the Gulf."

Lenaz served as the recreational representative on the Gulf States Marine Fisheries Commission's Mullet Task Force.

Headquartered in Ocean Springs, the Gulf States Marine Fisheries

(Opposite) Bas-relief on the Jackson County courthouse in Pascagoula honors the region's fishing heritage.

Commission was authorized by Congress in 1949 as a compact between the five Gulf states. Its charge was to "promote better management and utilization of marine resources in the Gulf of Mexico."

When it came to actual regulation, the commission—unlike its counterpart among the Atlantic states—mostly kept its head down and avoided the region's rough-and-tumble fish fights. The Gulf commission assumed a role that was more informational and advisory than regulatory: Staff generally collected data that experts at each of the state agencies produced, then compiled detailed reports and management plans that the regulators referred to in managing their respective fisheries.

The fishery commission's ten-member mullet task force included Kyle Spiller, a coastal biologist at the Texas Parks and Wildlife Department; Walter Keithly, Ph.D., a natural resource economist at Louisiana State University's Coastal Fisheries Institute; Harry Blanchet, biologist, Louisiana Department of Wildlife and Fisheries; Chris Dyer, Ph.D., an applied anthropologist and consultant with the National Oceanic and Atmospheric Administration; Henry Lazauski, Ph.D., certified fishery scientist, Alabama Department of Conservation and Natural Resources; Behzad Mahmoudi, Ph.D., marine biologist, Florida Marine Research Institute; Gene Raffield, president of seafood company Raffield Fisheries, Port St. Joe, Florida; Terry Bakker, assistant chief of law enforcement at the Mississippi Department of Wildlife, Fisheries and Parks; Mike Buchanan, biologist, Mississippi Department of Marine Resources; and the Mississippi CCA's Lenaz.

The task force was working on a Gulf-wide mullet management plan, which was not to be released for a year, in December 1995. However, by the time of the supervisors' meeting, the commission had reprimanded Lenaz for leaking portions of the report to the media in an effort to persuade the Ocean Springs City Council to endorse a net ban. Larry Simpson, executive director of the commission, felt compelled to warn the city council that "You can take all kinds of figures and distort them to say anything."

On their own turf and with their livelihoods on the line, net fishermen at the Pascagoula supervisors' meeting were less diplomatic. "I'd like to wrap those glasses around your neck," one told Lenaz.

From there, the meeting degenerated in tone until the supervisors had to call the sheriff to simmer things down.

Of the 17 people who testified, nearly all were opposed to CCA's proposal to curtail net fishing. Many talked about the number of jobs that would be lost—not just the fishermen's but those of the people who processed and distributed their harvest. Ultimately, they argued, the public would suffer by not being able to buy fresh fish to eat. Pascagoula fisherman Pete Floyd told the board: "The crisis they are portraying does not exist. Commercial fishermen are taking a select part of a renewable resource."

The supervisors ultimately voted 3-2 to delay any action on the sportsmen's request for a ban. The decision was not theirs to make, they said, it was the responsibility of the Commission on Marine Resources.

After the contentious Pascagoula meeting, Lenaz told the *Sun Herald*, "It appears somebody has turned the hostility notch up a little. Until now, there's been total disagreement, but it's been like a professional disagreement. For the first time, it's getting personal."

For commercial fishermen, the sportsmen's attacks had been "personal" from the beginning. Fishing was all some had ever known. They felt they'd hung in and hauled back so many nets that they were part net, and with families to feed, what was more "personal" than taking away their livelihoods?

To see Lenaz and other self-styled "conservationists" conspicuously circumventing scientific management and justifying their attacks with half-truths and pious save-the-fish rhetoric while dismissing net ban impacts on the fishing community frustrated and angered netters. Their moods weren't improved by the realization that this could be their last profitable roe mullet season.

BILOXI BACON

Coastal Conservation Association anglers had always obsessed over red drum and spotted seatrout, the inshore fish that they themselves most liked to hook. During the 1990s contagion, the group also homed in on a vegetarian species that rarely bit hooks and was the keystone of the net fishery.

The striped mullet was a traditional food fish on the Mississippi coast where it was marketed as "Biloxi bacon." The flesh had a nutty taste and an oiliness that lent itself perfectly to frying or smoking. Mullet was a high-quality yet affordable source of protein that was popular with consumers across the eastern Gulf and southeast Atlantic states.

Mississippi's catch was consumed locally and shipped east as far as Georgia where a handful of elite anglers had convinced state legislators to curtail most netting back in the 1950s. The fishery began to expand in the mid-1970s after export markets for the fish's roe were developed in Europe and, later, Asia.

During autumn, striped mullet bunched up in tight schools and migrated offshore to spawn. The quality of the fish's flesh wasn't at its best then, but the sacs of roe were plump and valuable. The short, intense spawning run could be lucrative for a skilled netter with the right gear.

After Florida and Alabama, the Magnolia State ranked third on the Gulf in landings during the 1960s and 1970s, then fourth during the 1980s and 1990s, after Louisiana fishermen began to develop that state's prodigious stocks. (Saltwater mullet could only be sold for bait in Texas.)

A lot of meals: This catch of mullet was taken in a haul seine that had smaller meshes than those in gillnets. *(Courtesy of Martin Young)*

Of the $860,000 that Mississippi's netters earned in 1994 from the sale of 15 species of fish, nearly half—$407,103—was attributable to mullet, most of which were taken during the roe run.

Historically, from 1950, the state's commercial landings ranged from a high of two million pounds in 1980—after purse seines briefly came into use—to 46,353 pounds in 1985, when several storms pounded the Gulf Coast from late summer through late autumn. Landings generally fell after 1980—with a decrease in purse seine use—and in the early 1990s varied from 803,253 pounds in 1990, to 439,334 in 1991, 474,334 in 1992, and 247,078 pounds in 1993.

According to the sporting conservationists, the drop in landings after 1990 was attributable to "overfishing" by netters that would, unless it was immediately stopped, lead to "commercial extinction" of the species.

The claim had plenty of resonance, given the saturation of Pew's "overfishing everywhere" media campaign; as for its relevance to the average angler, who was more interested in predatory redfish and trout, "Hundreds of thousands of pounds of mullet—a key food chain element—are being taken right now as we speak," Lenaz had told the *Sun Herald* in an article that appeared just after the November net-ban vote in Florida.

If sport fishermen weren't sufficiently incensed that the netters were starving their target species to death, Lenaz pointed out that they were also accidentally entangling the anglers' favorite fish as bycatch: "In those nets are a lot more than just mullet, too. There are specks, white trout, redfish, you name it." In fact, when compared with commercial landings, recreational fishermen caught mullet at about the same generous proportions they caught redfish and trout. According to NMFS data, recreational fishermen with cast nets took an estimated 2 million pounds of striped mullet in 1991, 1 million pounds in 1992, and 365,000 pounds in 1993.

HISTORICAL **COMMERCIAL** LANDINGS
OF STRIPED MULLET IN MISSISSIPPI

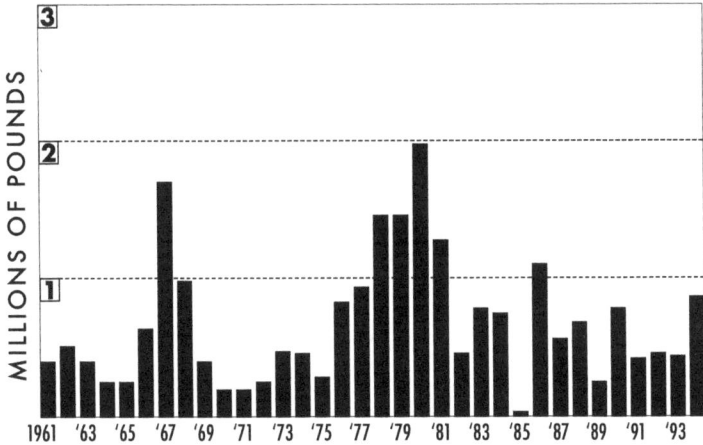

Commercial landings of mullet in Mississippi from 1961 through 1994. Sustainability's not a straight line—fish populations and landings vary with weather, regulations, environmental conditions, and other factors. The large catches in the late 1970s and early 1980s followed the introduction and brief use of purse seines by a Pascagoula fish dealer. Hurricanes in 1985 hampered fishing, then the 1986 harvest exceeded 1 million pounds. After regulators increased the minimum mesh size in the fishermen's gillnets in 1992 landings fell to 247,078 pounds in 1993, then increased as the fish grew larger, to 781,300 pounds in 1994, 615,142 pounds in 1995, and 842,186 pounds in 1996 (1995 and 1996 landings not shown in table).

(From "The Striped Mullet Fishery of the Gulf of Mexico, United States: A Regional Management Plan." Gulf States Marine Fisheries Commission)

Fear of overfishing led managers to increase the minimum mesh size in the gillnet fishermen's nets from 3 inches to 3 ½ inches in 1992, hence the drop-off in commercial landings. With a banana-shaped fish like mullet, half an inch makes a huge difference in the proportion of fish that escape to continue growing and spawn another day. And in fact, commercial landings did rebound, to 781,300 pounds in 1994, 615,142 in 1995, and 842,186 in 1996, the last year the state would allow fishermen to use their nets.

Mississippi biologists also observed that the drop in landings had roughly coincided with a low in the fish's natural eight-year population cycle. They noted, too, that their own testing indicated that there were plenty of juvenile mullet in the estuary.

Still, Mississippi trailed most other Gulf states in assessing its mullet stocks. Its biologists lacked sufficient data to accurately compute the spawning stock biomass ratio and spawning potential ratio, useful statistics to fishery managers. As a result, they didn't know precisely where the species stood in relation to the 30 percent spawning potential ratio that had become the accepted conservation standard for maintaining healthy fish stocks.

Nonetheless, state biologists agreed—and unhesitatingly stated in public—that the regulations then in effect were adequate for sustaining the mullet population.

Moreover, incidental take of sport fish in the mullet fishery was a nonissue, and the Gulf States Marine Fisheries Commission said as much: "Because mullet form tight schools during peak fishing periods (roe season), incidental catch of other species during this period is minimal."

Strike netters with a roe mullet set near the mouth of the Pascagoula River. An Ingalls Shipyard drydock is in the background.

COMMISSION ON MARINE RESOURCES DEFERS ACTION

Professional biologists in Mississippi and across the Gulf were frustrated that their scientific expertise was being eclipsed by the CCA's manipulative campaign rhetoric.

At the CMR's December 19 meeting, the Department of Marine Resource's Tom Van Devender, a biologist, briefed the commissioners with a reprise of the same presentation he'd made December 7 in Biloxi.

The biologist was "trying to show that there had been some comments that were circulating about that simply weren't true," he said, and hoped "to allay some of the fears that these comments had produced."

COASTAL EROSION PART ONE

On the topic of redfish, Van Devender flashed a graph before the commissioners: "You have a 42-year history of commercial red drum landings, and you see the sharp peak in 1978 which is the start of the purse seine fishery. You see the real low in 1990, that's when we put our 22-inch minimum size into effect. Other than that, you see very stable landings."

As for the 35,000-pound commercial quota, it was projected to be filled before Christmas, and the agency was issuing public notice of the impending closure.

"Just as a little sideline on this, a couple of weeks ago I received a telephone call from a gentleman who didn't identify himself at first. He said, 'When are you going to close your commercial red drum season?'

"I said, 'Well, we're tracking it, and I don't know, it's probably going to be before the end of December.'

"And then he identified himself as representing a red drum aquaculture facility in Louisiana. He said, 'Good, I'm glad to know, because we're going to put our fish on the market after you all close, and jack the price up. ... We just can't compete with these wild-caught fish.'"

Regarding mullet, Van Devender noted the peak landing years in the late 1970s, "with the advent of, again, purse seines in the Mississippi Sound." He also noted the 1985 low, which he attributed primarily to Hurricane Elena, which made landfall in Biloxi at the beginning of the run, but added that landings earlier that year had been "very stable."

However, he explained, landings data could be misleading absent explanations for the ups and downs. Fishery-independent data are essential. For example, he said, for the past 20 years, the Gulf Coast Research Lab had been netting for post-larval "young of the year" fish that were generally less than an inch long. "You can see," he said, "there is no overall decline, leading you then to believe that the spawning stock is very stable offshore. We followed that immediately with a juvenile index, again taken by the Gulf Coast Research Lab involving seines and other type fine-mesh nets in the marshes. Again, a 20-year data base, and you can see no decline."

Nonetheless, some people, "particularly one of our sportswriters," were declaring the mullet to be in decline, Van Devender said, "and it has caught the attention of the Gulf States Marine Fisheries Commission's mullet task force. In fact, the head of the task force, Dr. Behzad Mahmoudi in Florida, has written a letter to this particular sportswriter requesting that he recant what he has said and taking him to task for misinterpreting the data ... ."

People had also implied that spotted seatrout landings were down, said the biologist, but "We know that's not true. In fact, a week before this public hearing, James Warren here from the Research Lab ... had a workshop there at the Marine Ed Center, and I believe fishermen were in agreement that speckled trout landings ... for the past two or three years, they're better than they ever have been."

As he showed commissioners a chart comparing sport and commercial landings, Van Devender explained, "It's been true for a number of years that the recreational landings exceed commercial landings from several factors of degree. And, in fact, if there were to be problems in the population, we would certainly want to look at that segment of the fishery that's taking the most of them first," which elicited shouts and applause from the audience.

In response to sport propaganda that had the number of net fishermen going through the roof, Van Devender produced a chart that showed the sale of gillnet licenses in recent years to be "very stable." He concluded, "So this is the information that was presented at the public hearing, and again, the gist of it all is that we see in the landings data and in the fishery-independent data, no overall decline in fish stocks, particularly mullet, spotted seatrout and red drum, and that there is no reason, biologically, to put a ban on gillnets," which was met with even more enthusiastic applause.

On behalf of several fishing groups, SASI's Jean Williams told the commissioners, "We unanimously disagree with those who rush to close down a thriving industry. If indeed Florida's law stands—and we doubt that it will following that state's investigation into the fraudulent and deceptive way it was implemented—we request that this commission consider the following motion: We propose some type of license limitation on all saltwater gillnets, trammel nets, shrimp nets, purse-seine nets, and live-bait trawlers. In the event it becomes necessary to limit the number of these commercial licenses, it is absolutely imperative that a control date of December 19, 1994, be established."

Mississippi commercial fishermen wanted the invasion-from-out-of-state issue to be put to bed as soon as possible. Williams sought to do so by prohibiting new license sales after that evening. Commissioners briefly discussed a resolution to request the Legislature to establish such a control date, but because the item wasn't on their agenda, they took no action.

The panel also decided to defer further action on the net ban until after a second public hearing. Unlike the first one in Biloxi, which had concerned a possible ban in Harrison County only, this hearing wouldn't be in consideration of simply a localized ban, "because, frankly, to try to do one thing in one county, I believe we all would agree, would be a disaster, and it would not be worth it," explained Commission Chairman Sherman Muths.

Instead, Muths proposed that the next hearing take place in a location "where everybody can get in," and it would air whether or not gillnets should be banned in *all* the marine waters of the state. The commissioners could then "do anything from a total ban to doing nothing to something in between. We have total flexibility of doing that."

Nobody applauded Muth's proposal, which the panel approved. The next public hearing on the issue would be held in the Biloxi Coliseum on the evening of January 11.

SENSELESS SLAUGHTER

To drum up attendance at the January 11 hearing, GCCA lit up anglers with an ad in the Sunday, January 8, edition of the *Sun Herald*. Over a photograph of a redfish that was simply being landed in a net, the bold caption asked, **"Will you help stop this senseless slaughter?"**

The text of the ad continued:

"You're looking at a killer. It's a gill net. There are forty miles of gill nets being used on the Mississippi Gulf Cost (sic) today. These nets won't stop killing fish until the last fish is caught. Unless we do something to ban the nets.

Gill Nets—Deadly, Effective, and Illegal Elsewhere

Gill-nets are usually set to catch mullet, valuable because their eggs, or 'roe', are in big demand in oriental markets. It's bad enough that processors strip out the roe and discard the mullet flesh or use it as crab bait. But, once these gill nets are put in the water, they blindly dispense death.

Tens of thousands of speckled trout and redfish die in gill-nets each year. Marine birds, attracted to the nets, are often fatally entangled. And, sadly, cases of our beloved Atlantic Bottlenose Dolphin drowning in these nets are well documented.

Here in the South, gill nets are illegal in Texas, South Carolina and Georgia. Just recently, 72% of the citizens of

Florida voted to ban gill nets from thier (sic) waters. Alabama and Louisiana are expected to ban the nets this year. So, will Mississippi be the target for the 1,000 miles of Florida gill-neters (sic) the voters put out of business? For the sake of our children, let's hope not.

Is this an end to commercial fishing?

No one wants to ban commercial fishing. Just the deadly gill nets. Very few of the fish on local menus and in local markets are caught by gill nets. Other forms of commercial fishing, such as commercial hook-and-line fishing, are much more selective and much less impactful. They lend themselves to regulation and management. The mass slaughter you get with gill nets simply cannot be managed. Banning the nets is the only choice.

The 164 people using gillnets say they have a right to make a living. We can sympathize with them. But we cannot allow the needs of 164 gillnetters to destroy a resource that belongs to all Mississippians.

On Wednesday, January 11, you can do something about it.

The Harrison County Board of Supervisors and the City Councils of Ocean Springs, Biloxi, Gulfport, Long Beach, Pass Christian, and Bay St. Louis have all passed resolutions backing the ban on nets. But you're probably wondering what you can do.

The Missisippi (sic) Department of Marine Resouces (sic) (DMR) will hold a final public hearing on Wednesday, January 11, at 6 p.m. at the Mississippi Coast Coliseum. The DMR has the power to ban the nets. You can be sure the commercial fishing interests will be there in force. Make sure you are there, and help stop gillnets from slaughtering our sealife."

DEATH BY A THOUSAND HEARINGS

BIG GUNS

On January 10, in neighboring Alabama, commercial fishermen and their supporters marched in downtown Mobile; they were protesting the sportsmen's proposed net ban and "discrimination against seafood people."

On the evening of January 11, Mississippi's Department of Marine Fisheries held its second and final public hearing on the sportsmen's proposed net ban. The venue this time was Biloxi's Mississippi Coast Coliseum which could hold upward of 11,500 people.

After DMR staff yet again presented data showing that fish stocks were healthy, the floor was opened to public testimony.

Biloxi and sister city Gulfport were the home ports for most of the state's charter-boat fleet. Many of the inshore guides ran their customers over to Louisiana to fish. Those with larger vessels, such as Air Force retiree Tom Becker, fished offshore for red snapper and other species, none of which were caught with nets. Still, in his capacity as president of the Mississippi Charter Boat Captains' Association, Becker asked the commissioners to ban the commercial fishermen's gear.

So did the American Sportfishing Association's vice president of government affairs, Norville Prosser.

Headquartered in Washington, D.C., the ASA was the product of a recent merger between the Sport Fishing Institute and the American Fishing Tackle Manufacturers Association. Now, as the corporate recreational fishing industry's advocacy group, the ASA had a hand in coordinating the national attack on the seafood industry.

Prosser, who was touring the embattled Gulf Coast states at the time, urged the commission to consider the economics of sport fishing in its decision. More than 100,000 resident and 60,000 nonresident anglers made more than 670,000 fishing trips to the coast each year, he said. Those saltwater anglers contributed more than $71 million annually to the coastal economy by spending money at local bait shops, gas stations, motels, restaurants, and other small businesses. Add in freshwater fishing near the coast and the figure jumped to $125 million; statewide, the sum approached $500 million, he said.

"The economic health of this industry is linked to a well-managed and vibrant fishery," said Prosser, insinuating that with nets, the fishery could be neither: While no obvious trends in spotted seatrout landings were then evident, the mullet catch showed recent decline, and the red drum harvest had declined significantly since the early 1980s, he said.

As for the net ban itself, Prosser noted that the Department of Marine Resources, "like many other marine fisheries regulatory agencies, is engaged in debate on the best practical choices to manage publicly owned renewable fishery resources in ways that afford resource protection and guarantee the greatest social and economic benefits to citizens.

"To that end, the American Sportfishing Association joins with the Department of Wildlife, Fisheries and Parks and the growing list of concerned coastal city and county governments in supporting this ban."

Of the 930 people that signed in at the hearing, 899 expressed an opinion on a net ban: 544 were opposed, 355 in favor.

GOOD PRESS/BAD PRESS

With the public hearings concluded, the Commission on Marine Resources was set to take final action at its January 24 meeting. Meanwhile, both of the state's coastal papers weighed in on the issue in their January 15 Sunday editions.

The caption of the editorial in Jackson County's *Mississippi Press* needed little elaboration: "Scientific data has the leverage: Don't ban gillnets."

The issue was "highly emotional" on both sides of the argument, said the paper, but it was time to put aside the emotion and look at the facts which are "clearly on the side of the gillnet fishermen."

With data and quotes from biologists at the Department of Marine Resources, Gulf Coast Research Laboratory and Gulf States Marine Fisheries Commission, the paper made it clear that fish populations were fine, with nets in the water. Not only was a ban unwarranted on scientific grounds, but it would deprive "non-fishing consumers, including tourists who eat in restaurants" of "a plentiful and reasonably priced supply of seafood."

Sport and commercial fishermen could continue to coexist in Mississippi Sound, ventured the editors, although, to that end, "Perhaps some tighter restrictions should be imposed on gillnet fishermen … ."

The editorial concluded by encouraging the CMR to "base its decision on … facts rather than emotion or the numbers of fishermen involved on each side of the issue. Fish stock is carefully monitored. If there comes a time when scientific data indicates a fish species is threatened, that will be the time to ban gillnets."

Even if the Pascagoula-based *Mississippi Press*'s commentary fell flat at the end, fishermen found its high-road tone a welcome counterpoint to coverage by the paper's rival on the coast, the Biloxi-based *Sun Herald*.

In "Sportfishing means big bucks to Coast, state economies," the *Sun Herald*'s outdoor writer John Lambeth reported on the recent visit by sport-fishing-industry executive Norville Prosser.

The article reiterated Prosser's claims of recreational fishing's outsized economic impacts, and his disappointment "'in the quality and quantity of scientific, fishery-independent monitoring data available to us for making an assessment of the condition of finfish stocks.'"

Absent any balancing viewpoints, the article appeared to support the net ban's proponents, with their repeated claims of socio-economic superiority and biological ambiguity, the latter of which jibed with the "precautionary principle" advocated by environmental nonprofits—the idea that the environment should be protected from plausible if not proven risk.

Elsewhere in the same edition of the newspaper, in his outdoors column titled, "Net ban question is not a complex issue when viewed as it should be," Lambeth predicted a landslide vote in favor of banning nets.

The decision would probably not be based on any scientific data, he reasoned, because science, "through its own lack of hard data," had become a "moot point." What little science there was, was "in great dispute, even among some scientists," which was "confusing and regrettable, but not a problem," because the commission was charged with "protecting the resources *and* serving the public."

One thing couldn't be disputed, "if nets *are* removed, it will *not* harm the resource," which left only the "will of the people" for the commissioners to consider, Lambeth reasoned. And on that basis, "The vote should be 5-2 for the ban, with no compromises."

His prediction hinged on the premise that six of the seven commission members were appointed as representatives of designated "user groups," and would vote accordingly.

For example, Henry Boardman, of Gulfport, occupied the "charter boat operator" slot. The for-hire fleet was on record in support of a ban so, one vote in favor.

Gulfport attorney Sherman Muths represented private recreational anglers. "There's no doubt about where the overwhelming majority of sportfishermen stand on this issue," so, another vote for a ban.

Pascagoula's Doug Horn filled the commission's "commercial seafood processor" slot, and Oliver Sahuque, of Lakeshore, in Hancock County, was the panel's "commercial fisherman." As seafood producers, the two would naturally vote against a gear ban, thereby lodging the only two contrary votes.

William Mitchell, the "non-seafood industry" appointee, would be "expected to vote with the majority of the public since there would be no adversity heaped on the resource."

Vernon Asper of Diamondhead, in Hancock County, represented environmental interests: "At last report, virtually every Coast environmental group supports the ban. Accepting that, Asper would have no other choice but to vote for a ban."

The seventh member, Webb Lee, sat on the coastal commission by virtue of his position on the Commission on Wildlife, Fisheries and Parks: "Lee, too, is bound to vote with the will of the people—an easy task, since the Department of Wildlife, Fisheries and Parks is strongly supporting banning gillnets in Mississippi waters.

"So, where's the question? There should be none."

CMR TAKES ACTION

On January 24, 1995, before an edgy crowd at Biloxi's Convention Center, Commissioner Webb Lee moved for a "total ban on the use and possession of gill and trammel nets in the marine waters in the state of Mississippi effective May 1, 1995." The fishermen's worst nightmare was here.

Lee's motion was seconded by charter-boat captain Boardman but before it was put to a vote, the panel's environmental representative offered an alternative.

In a lengthy preamble, Asper reminded his fellow panelists that the commission was established to manage the resources "for all the people of the State of Mississippi." Its members were charged with doing that "in the most objective way possible," and since they'd been appointed rather than elected, could act "in the absence of any kind of outside pressure."

With a Ph.D. in Marine Geology and Geophysics from the Woods Hole Oceanographic Institute, Vernon Asper was a professor of marine science at the University of Southern Mississippi and an extreme oceanographer whose research on particle dynamics took him around the world, including to Antarctica. "Data are basically my life," he said, so it would be very difficult for him to turn his back on other scientists and their data which, of course, did not support taking the fishermen's gear for *biological* reasons.

Asper reminded the commissioners that they had a "serious issue" before them, with "a lot of momentum," and cautioned that if it weren't dealt with properly, "contention" and "strife" would escalate. His solution was a good compromise, which he defined as one in which "everybody feels like they gave up a little more than they got," or "you feel that you really don't like it but because of the situation you have to live with it."

The "situation" demanded that he "seriously restrict" net fishing, said Asper, and after conceding that he felt "a little bit bad" and was "almost embarrassed" to do so, because he didn't feel it was "even a serious enough problem to warrant this drastic a measure," he offered the specifics of his proposal.

To separate sport and commercial fishermen, who were "getting into each other's face," net fishermen along the more populated areas of the coast would be moved further offshore during the daytime, "when most of the recreational fishermen are out there." In the nighttime, "they can come in closer."

To further separate the user groups, netting would be prohibited on weekends and holidays. "Now, I realize that this is an amazing thing to ask commercial fishermen to do," said Asper, as he appealed to that segment of the audience. "I realize how important it is for you … especially in roe mullet season … to be out there when the fish move out … And I realize that this imposition I'm laying on you is going to be tough to swallow," but, "given the situation, we need to do something."

To address the problem, or "the perception anyway," that fishermen were curtaining off an entire area with set-nets "end to end overlapping so that not a single fish escapes," Asper proposed that nets be set no closer to one another than the length of a net. "What that allows is a space for fish to get through."

While on the subject of anchored nets, he also encouraged setnetters to comply with existing law, which vaguely required them to attend their gear "continuously from a distance which is not too great," to avoid the "perception, anyway, that these nets are out there fishing by themselves, and the fishermen are not around."

Asper also addressed the enforcement division's complaint that its agents were unable to ticket violators found in closed waters unless they were caught with their gear in the water. Henceforth, anyone in closed areas would be in violation if they had a fish in contact with a net, even if that net was onboard.

So far, Asper's proposals, particularly his restrictions on when and where fishermen could operate, were crippling yet didn't completely wipe out net fishing. Then the oceanographer went off on a tangent.

"If a gillnet is lost or it's abandoned or for whatever reason it gets away and floats out unattended or sinks to the bottom or disappears into water deeper than it floats, it's going to be fishing forever, forever, and ever. And from the environmental perspective, regardless of what other problems there may be with gillnets, that issue alone is probably the most grievous because having a gillnet out there which lasts forever is really a serious problem."

Biodegradable nets, on the other hand, would not only eliminate the "ghost fishing problem" but would be less efficient tools because, "unless we come up with an amazing new substance," cotton or linen nets, even if dipped with a protective coating, would not be as easy to fish as nylon nets. "So we are going to allow a certain amount of inefficiency into the system, which returns us to the olden days but it eliminates a couple of big problems."

The incidence of "ghost nets" fishing "forever and ever" was more prevalent within the literature of environmental nonprofits like Greenpeace and Asper's Sierra Club than it was in the waters of Mississippi Sound. Which is not to say that factory ships weren't losing—or discarding—tons of nets and trawls on the high seas. But these were the vessels that, according to a favorite claim by environmental campaigners in the 1990s, pulled trawls immense enough to engulf jumbo jets.

Fishermen in Mississippi's inshore fishery weren't operating vessels the length of football fields, they were fishing from skiffs the size of pickup trucks.

The components in their nets were expensive to purchase and even more time consuming to assemble. Nobody wanted to lose theirs. And discarding worn-out webbing in shallow waters would likely foul somebody's propeller, possibly their own!

In strike netting, where fishermen kept in contact with their gear, losing a net was impossible; in set-netting, where the gear soaked while it intercepted passing fish, it was at least possible.

At the time, there were no qualifications for purchasing a net license and no formal training was required. In a storm, an insufficiently anchored net could get away from a novice. Or an outlaw—and Mississippi's fishery had its share—could abandon his if things got too hot when fishing in closed waters.

Then what? A gillnet used in shallow waters tended bottom. If lost, it would drift with the current until the leadline hung up on some obstruction and the ends would be swept together into a tight spiral.

At the same time, a moving net's webbing would entangle oyster shells, twigs, and other debris, pulling the buoyant corkline down and the leadline up until the whole thing was rolled up like a rope.

In short order, a net lost in the Sound would be reduced to a pile of glop that in the warm waters of the Gulf, where clean substrates were at a premium, would soon be encrusted with barnacles, oysters, and other reef-building sea life.

There were no "ghost nets" haunting Mississippi Sound, standing erect and catching fish forever and ever, and the commissioners weren't presented with any data indicating that there were.

Asper, however, was under the gun, trying to thread a needle at a time when the animosity toward commercial fishing in general and nets in particular must have astonished the academic. Though he'd told his fellows that they were able to act "in the absence of any kind of outside pressure," the reality was that the sportsmen were already threatening to have the Legislature abolish the new coastal panel if they didn't get their way.

As he coaxed his fellow commissioners to approve his plan, Asper told them that a thumbs-up would allow them to address more important issues. "We have, so to speak, bigger fish to fry than nets. There are things out there in the environment which are seriously threatening our resources," such as wetland, pollution, and development issues, he said. "We're focusing on a little tiny part of a much, much bigger problem and, especially, we're fighting amongst ourselves when we should not be doing that. We should be getting along. We should be agreeing that we have a problem area out there and that is that we are looking at a dwindling resource because of the increasing development as well as all these other problems. We need to be working together on that. We need to get past this issue."

Give his program a year to work, he urged, and if there was a problem after that, "I will lead the effort to get back in this room ... and we will ban the gillnets. I promise you we will do that."

Conceding again that he wasn't "convinced that we have such a serious problem that we need to resort to a drastic measure like a total ban," Asper made the same pledge to the state's "environmental and conservation groups," asking in return that, during the one-year trial period, they refrain from "taking it to the next step," by going around the commission and appealing directly to legislators and the governor.

Commissioner Lee called for a vote on his original motion to ban nets across the coast. The motion failed with three votes for—by Lee, Boardman, and Muths—versus four votes against, which included the panel's seafood people, environmentalist Asper and Pascagoula auto dealer Mitchell.

Before considering Asper's compromise, Lee urged the panel to vote it down because it presented "an enforcement nightmare" with its prohibitions on netting in some locations but not others. Asper countered that there would be enforcement problems, no matter what they did. The *easiest* thing, for enforcement, would be to make it completely illegal to possess a net, but that didn't mean it was the *right* thing to do, he said.

In addition to his own concrete proposals to facilitate enforcement, Asper assured his peers that the commercial community was eager to cooperate: "I think these people feel that they have been represented and we're really looking out for their interest and they're going to help us make it work. And, because we are going to be giving it one year, they understand how important it is to make this thing work."

The panel approved Asper's package by a 4:3 vote. Once again, Lee, Boardman, and Muths comprised the minority.

Under the new rules, from Hancock County's Bayou Caddy, east across all of Harrison County to Ocean Springs, just inside Jackson County, netters had to remain at least half a mile from the shoreline from 6 a.m. to 6 p.m., and at least a quarter mile from shore from 6 p.m. to 6 a.m. Essentially all of Jackson County's waters were exempted from that provision, but it was unlawful to use a net or have a net in contact with fish in a boat in *any* marine waters of the state, between 6 a.m. on Saturday mornings and 6 p.m. on Sunday evenings, between 6 a.m. and 6 p.m. on legal holidays, and at any time in the upper reaches of St. Louis, Biloxi and Pascagoula Bays. No gill or trammel net could be set within a quarter mile of another net, and on or before January 1, 1997, all gillnets and trammel nets had to be constructed of an approved biodegradable material.

TWO MORE YEARS OF CONFLICT

Most of the Commission on Marine Resources' new restrictions on when and where fishermen could deploy their gear were to take effect quickly, in April 1995. As for the biodegradable mandate, netters didn't have to comply until January 1997. It wouldn't be a peaceful two years.

Commercial fishermen chafed at the so-called "compromise," because, they said, it was excessively restrictive, unnecessary, and inequitable.

"Why must one user group which only accounts for 20 percent of the catch take 100 percent of the responsibility?" asked SASI's Jean Williams.

Williams acknowledged that four of the commissioners "spent many agonizing hours trying to find something that everyone could be satisfied with. We appreciate what they did," but the end result was unfair, she said. "They have put a lot of severe restrictions on a group of people who don't deserve it. Scientific data says there is no reason for this.

"We have a big problem with the ban on net fishing on weekends and holidays," she added. "If we have a northerner on a weekend, the roe mullet will be gone before anyone can catch them. We can't tell the fish to run only during the week."

Third generation fisherman Eli Ross of Biloxi considered the weekend closure to be the most damaging of the provisions. Being pushed at least half a mile off the coast in the daytime wasn't as damaging, he said, because Biloxi's touristy manmade beach was "where you've got all your jet skiers and speedboats, so we don't catch fish during the day there anyway.

But what hurt us is that we can't work on weekends no more—just so the tourists and sportsmen don't have to see us ugly fishermen. What I don't get is how can they tell us we can't work on weekends? We don't tell them they can't sport fish on weekdays."

Williams said that fishermen were willing to comply with the requirement that their gear be made of biodegradable material but asked why the same restriction wasn't placed on sport fishermen.

Indeed, if each of Mississippi's estimated 160,000 resident and non-resident saltwater anglers were equipped with a single spinning reel filled with 250 yards of monofilament, they would be carrying nearly 23,000 miles of plastic line into the Sound, enough to stretch from Jackson, Mississippi, to Los Angeles, California, more than a dozen times.

Given the political realities of the time, Mississippi's commercial fishermen were hardly positioned to steer management toward the equitable application of regulations. They were lucky to have survived and knew it.

"After Florida, they thought, 'We'll just steamroll right on across the Gulf.' Everybody was talkin' like it was a done deal, and we were dead meat," said Hancock County fisherman Bob Metz. "But it just didn't work out that way."

Save America's Seafood Industry's Jean Williams knew the fight wasn't over. "Although we dodged a bullet this time, the issue won't go away," she said. "They will be back."

SPORT FISHERMEN REALLY UNHAPPY

Sportsmen hadn't given up anything in the "compromise" except the thrill of winning a complete victory which of course meant no nets anywhere.

"Our side lost," said Ray Lenaz. "What they've done is move the perception of the problem—move the gillnetting from the beaches so people don't see it. They still have not solved the problem."

"This is an affront to all of us," added GCCA president Terry Waldrop, of Gulfport. "They think they can buy us off with this proposal."

Asper's attempt at compromise was "absurd," said Waldrop. "It just polarized the recreational fishermen even further." If any compromise should have been adopted, he said, it should have been the one that GCCA proposed which, in lieu of a total ban, would have allowed net fishing only during the three-month roe mullet season, and then not on nights or weekends.

"The new regulations do not address the problem of out-of-state gillnetters, there are no stricter penalties and there is little way to enforce the measures," Waldrop complained. "This is the same system we have now."

Not only would the GCCA ignore Asper's plea to give his plan a chance, but the group would fight the commissioner's formal confirmation by the state Senate, said Waldrop, a manager at an industrial construction company.

GCCA would also try to convince the Legislature to enact its own net ban, and if that should fail, said Waldrop, his group would pursue the nuclear option: the state's newly approved referendum process to put the netting question to a vote by the public.

VOTERS TO THE RESCUE? MAYBE NOT

Mississippi had only recently adopted the ballot initiative, for a second time. The process had formerly been allowed, until 1922, when the state's Supreme Court struck it from the constitution on a legal technicality.

Initiatives and referendums had been widely discussed in campaigns for the 1991 fall elections, and the 1992 Legislature put the issue on the ballot. Hailed as a progressive reform of government, it was approved by 70 percent of voters in the 1992 fall elections, making the state the most recent to adopt the procedure.

Still, it wasn't as easy to get an issue on the ballot in the Magnolia State as it was in, say, California or Florida. In those states—where recreational fishing interests had recently persuaded voters to nix nets—special interests had only to gather a required threshold of signatures on a petition.

The process was less direct in Mississippi where a proposal still had to be approved by the Legislature before it could be put on the ballot.

On January 25, the day after the coastal commission approved its compromise package, legislators brought up a bill calling for a statewide referendum on nets.

Gulfport's Sen. Billy Hewes, chairman of the Senate Ports and Marine Resources Committee, wanted to discuss the proposal among committee members without calling it up formally. A common practice, the informal exchange helps legislators craft their amendments.

Jackson County's Sen. Brad Lott, however, insisted on calling the measure up formally. After Hewes reluctantly complied, Lott got up and walked out of the meeting. His hasty exit left the committee without a quorum which, under legislative rules, killed the bill.

The maneuver enraged sportsmen.

BACK TO JACKSON

The Mississippi Legislature had established the Commission on Marine Resources to manage coastal affairs, but the body had by no means relinquished its own authority to do so.

On February 8, the Senate passed a bill banning all finfish nets within one mile of the Mississippi coast, including the shorelines of Jackson County and all the state's barrier islands.

If passed by the House, SB 2591 would erase the commission's compromise package of regulations. While the compromise had applied only to gill and trammel nets, this bill included purse seines which, in addition to their use in the menhaden fishery, were also allowable in Mississippi Sound for the harvest of food fish like black drum.

In arguing for his bill, Biloxi's Sen. Tommy Gollott said, "Mullet, because they spawn so close to the surface, are being caught in droves. These people with gillnets and trammel nets can encircle that whole school and catch about everything in the water."

Gollott's bill would push fishermen into waters that were too deep for their nets to be effective, prohibit them from netting on weekends and holidays and from harvesting any mullet during the first half of the roe season. It also prohibited individuals from selling their catches without a commercial license, increased the fine on a first offense for illegal commercial fishing from $2,000 to $5,000, and allowed authorities to seize the offender's boat, motor, and equipment. The maximum fine on subsequent offenses was $10,000, and forfeiture.

SB 2591 banned netting in all coastal bays and bayous, and all commercial fishing except oystering in marine waters north of coastal U.S. 90.

The latter closure had been proposed by Vernon Asper at the commission's January 24 meeting. Intended to protect the inland nursery areas, his measure targeted netters, crab trappers, and particularly, the bait shrimpers who supplied 41 live-bait camps. Trawling in the nurseries was "one of the most destructive practices we've got going," Asper had said. The CMR had scheduled a round of public hearings on the matter which, coincidentally, were being conducted in the same week that the Senate was considering the issue.

Such dual regulation chafed commission chairman Muths: "It makes no sense to have people from all over the state voting on how to manage marine resources," he said. "You can't have marine resources co-managed by the commission and the Legislature. That'll never get us into the 21st century."

Ironically, Gollott, the senator who sought to override the commission with his own legislation, had also sponsored the bill that created the commission the previous year.

"This is the Legislature reaffirming our position," said Gollott. "It's us doing our legislative duty in what we think the people on the Mississippi Gulf Coast, especially Harrison County, want to see."

His bill had passed the Senate by a 47-2 vote, with coastal Sens. Brad Lott of Pascagoula and Tommy Robertson of Moss Point in opposition.

"That has gone way too far," Lott said. "There's no imaginary line in the ocean a mile from the shore. The shore bends back and forth. The law enforcement people told me it was unenforceable."

Fighting on two fronts, fishermen found themselves again trekking up to Jackson to lobby against this and other damaging legislation.

"Senator Gollott wanted all the bays and bayous closed," explained Jean Williams. "We argued with him that we could accept above Highway 90 but nothing more. But what they were wanting, of course, was to close the Sound down. They didn't get that—we got that thrown out.

"We supported higher fines if you were caught in the estuary above Highway 90; we strongly backed real high fines and taking your equipment on the second offense. But we didn't want the fines to be that much out front [in the Sound] because these guys don't have LORANs, etc. —they could be in the wrong place at the wrong time, and they would be in trouble. If they did implement that out there, we'd insist on physical markers. So we got all that thrown out."

After the House Conservation Committee amended SB 2591 to mirror the package that had been passed by the Mississippi Coastal Commission, the committee's chairman declared that there was no reason to duplicate what the commission could already do. With no further House action, the bill died in early March.

TWO MORE YEARS OF CONFLICT

ACCORDING TO WHOM?

As GCCA continued its campaign to stop net fishing, the group targeted dining establishments. Members distributed cards to local restaurants with estimates of their annual dining expenditures along with the message, "The customer who left this card requests that you join in the responsible use of our marine resources and believes your fish can be supplied by hook and line commercial fishermen. Please do not serve net caught fish."

Meanwhile, the commercial industry educated seafood consumers themselves with an assortment of fact sheets and other materials.

One of SASI's handouts, "Save Our Seafood: Questions and Answers about the Campaign to Ban Commercial Net Fishing," succinctly countered each of the sportsmen's arguments, and in a single sentence distilled the issue to its essence: "It is not possible to provide low-cost fish without nets."

The brochure included several caveats for the unwary consumer of seafood *and* media:

> When you see pictures of nets being improperly used, realize the campaign to ban nets is a highly partisan political campaign. Those behind it use all the techniques of political propaganda. Ask where the pictures came from and when they were taken.
>
> When you read something in the media or watch television, ask whether the information applies to Mississippi waters.
>
> When you read your newspaper, remember, 'Sports' writers advocate 'Sport' fishing. Often, they are columnists, not reporters, and so they may not be fair or objective. They often go on fishing trips paid for by equipment companies, some even own charter boats.
>
> Finally, ask: Does the information cite objective scientific study? Or does it merely express feeling and emotion?"

THEY'RE GREEDY, WE'RE NOT

When SASI's informational brochure cautioned the public that outdoor columnists "may not be fair or objective," and might "even own charter boats," the group had someone in mind.

John Lambeth ran a charter boat out of Gulfport. He also wrote articles and columns for the Biloxi-based *Sun Herald*, which reached a wider audience on Mississippi's coast than any other publication. Lambeth had been with the paper since the early 1980s and as "Outdoors" editor he'd been irking commercial fishermen for years.

Given the mood of the time, they had found a June 1993 article conflating them with yesterday's "greedy" market hunters beyond annoying.

The piece, titled "A modern day buffalo hunt?" was a riff on a press release that had been widely distributed to national media by the Marine Fish Conservation Network, a newbie coalition of environmental nonprofits that was trying to reform federal fishery management.

"This is the story of the buffalo hunter on the Great Plains of the Old West," it began. Lambeth went on to outline the history of America's terrestrial wildlife management or, at least, the narrative that had been repeated for more than a century.

It began, in Lambeth's telling, in the elysian days of abundance when thundering herds of bison roamed the west. Then "professional buffalo hunters" began to kill the animals for their meat and hides. When the "last buffalo fell," the buffalo hunter "turned to bear, to elk, to moose, to deer. And, one by one, he nearly sent each of these critters down the path of the buffalo—to commercial extinction.

"Commercial extinction is a term used to describe a resource no longer available in numbers to support commercial, or any other, harvest. In other words, the resource has been effectively used up.

"During the unbridled slaughter of land animals came a bit of foresightedness that has halted the greed-prompted killing of wild animals for commercial purposes. Public disgust at the incredible slaughter of beast and fowl and a growing conservation ethic shortly after the turn of this century brought about the formation of many state game and fish agencies and the subsequent regulations we now live under.

"The market hunter literally shot and killed his way out of business."

There was, of course, never a time in our nation's history when "market hunters" were the only people killing our wildlife. From the first arrival of men to North America every type of animal—from groundhogs to grizzly bears—was killed, without regulation, by everyone who needed food for subsistence, hides or furs for warmth, or protection from dangerous predators or crop-destroying varmints. They were also killed for sport.

That included bison, which were hardly managed for sustainability. They were intentionally decimated as a matter of national policy driven by western expansionists.

According to "The Buffalo" author Francis Haines, "Army officers looked at the slaughter and approved, for once the herds were gone the Plains Indians would be more peaceful. The cattlemen were pleased too; every buffalo killed made room for one more longhorn steer. The farmers came west in droves, eager to plant wheat once the fields were safe from the trampling herds."

Because commercial activity typically leaves a paper trail, it's much easier to quantify the wildlife that's marketed versus that which is killed for subsistence or sport. However, if some objective historian were able to compare the relative impacts that these activities had on our land-based resources, before they were regulated, he or she might be as surprised as inquiring folks are today when they learn of the recreationists' outsized impact on our marine resources.

Commercial fishermen of course already knew that, and seethed when

Lambeth wrote, "Many conservationists see another buffalo hunter in operation today," and insinuated that a cascade of calamities was attributable solely to commercial fishing:

"Two stocks of Pacific salmon are extinct, two stocks are declared endangered and five more are awaiting endangered status; some 65 species of commercially profitable fish are declared overfished and landings of giant bluefin tuna are down 90 percent; sharks, redfish, king and Spanish mackerel, grouper, snapper, trout and more are among the overfished or are commercially extinct."

Dams, of course, were the salmon's worst enemy. And most of those other fish—at least those in the Gulf of Mexico— were being hammered harder by recreational fishermen than commercial. Yet, to save the fish, a coalition of "high-powered conservation and environmental groups that, heretofore, have been concerned with land-based threats to the environment, have banded to form the Marine Fish Conservation Network," the article/release announced.

In fact, those groups had just the week before participated in a workshop in St. Petersburg, Florida. They included the Sierra Club, National Coalition for Marine Conservation, Center for Marine Conservation, National Wildlife Federation, National Audubon Society, World Wildlife Fund and Greenpeace. The recreational Gulf Coast Conservation Association and its affiliate, the Florida Conservation Association had also sent representatives to the workshop, and so were fully apprised of what was to come:

The "nonpartisan" coalition's members were formulating "public education plans" that would ensure their victory in Congress. "We don't have time to fight a war, because if we do, the fish will disappear while we're fighting," said David Allison, of the Center for Marine Conservation. "We have to show and convince everyone that conservation is critical right now."

Lambeth's piece was illustrated with a photograph of a few Mississippi shrimp trawlers, heading out to work. The photo was captioned, "Commercial fishermen of today have been likened to the buffalo hunter of the Old West."

Enraged commercial fishermen responded to the article with a barrage of letters-to-the-editor that attempted to correct some of its glaring omissions—many pointed out that the fisheries cited in the article included not just commercial fishermen but private recreational anglers and those hybrid fishermen who got paid to take recreational anglers fishing—charter-boat captains like Lambeth himself.

Although fishermen would later complain that the *Sun Herald* was selectively editing or even discarding their letters, the paper printed enough of them to elicit a counter-campaign by GCCA:

"Several letters to the Editor of *The Sun-Herald* have attacked sportfishermen as the rapists of the Gulf. We must write replies in our own defense," prodded the group's newsletter. "If we, GCCA and sportfishermen, do not answer the call, we deserve exactly what we get: NO MORE ACTION FROM THE LEGISLATURE AND NO MORE RESPECT FROM THE GENERAL PUBLIC. If we want gamefish status for trout and redfish and a ban on gillnets, we must have the public on our side. ... Our sort [sic] deserves your time and effort, and John Lambeth, who has been a loyal spokesperson for conservation, legislation, GCCA and sportfishing deserves to have our support in print."

HITTING THE STREETS

On Friday, February 10, 1995, between trips to Jackson, where Save America's Seafood Industry was trying to control damage in the Legislature, the group picketed the *Sun Herald* in Biloxi. Though precipitated by a recent editorial, the protest was intended to draw public attention to what the marchers claimed was the paper's ongoing unfair treatment.

"The abuse of media influence, outright, blatant discrimination, repeated printing of lies concerning our industry on the letters page while our letters are not printed, or are 'edited for content,' has prompted this protest," said SASI's communications director Hilton Floyd.

"People believe what they read, and the articles printed in the *Sun Herald* clearly show a biased opinion against the commercial fishing industry," added Jean Williams.

About 30 SASI members marched outside the newspaper for nearly two hours. They carried signs with slogans such as, "The Sun Herald Discriminates Against ME!" "SUN HERALD RUN BY GCCA," "The Sun Herald promotes propaganda," and "Biologists don't lie, The Sun Herald does."

The picketers criticized the paper's coverage of the redfish, red snapper, and shrimp fisheries over the previous decade, yet its recent editorials were "the icing on the cake," Williams told a journalist, who had the awkward task of reporting on a protest of his own employer.

"The *Sun Herald's* editorials supported the gill-net ban in spite of all the data provided by the scientific community. But if scientific data backed up the claims you've made, we wouldn't be here," said Williams, presumably because there wouldn't be any fish left to net.

Protesters objected particularly to Lambeth's work which they said was reflective of the writer's membership in the partisan GCCA. They also complained that their letters-to-the-editor were being screened and noted that the editorial pages were overseen by Lambeth's wife who, the paper

(Opposite) Piling it on: "Wasted Fish Stink! So does the politics as usual that threatens fishing in the Gulf of Mexico." This ad, by the Marine Fish Conservation Network, appeared in Mississippi newspapers in May 1995, after the U.S. House Natural Resources Committee failed to approve legislation backed by the two-year-old network.

Wasted Fish Stink!

So does the politics as usual that threatens fishing in the Gulf of Mexico.

Each year more than one billion pounds of recreationally and commercially valuable fish are killed by indiscriminate fishing.

Shrimp trawlers annually kill and waste:

- **34 million red snapper**
- **5 million Spanish mackerel**
- **650,000 king mackerel**

We must eliminate this waste in order to keep the fishing industry viable and to ensure you and your children will be able to catch fish in the Gulf.

The Future of Fishing in the Gulf Depends on You!

The House Resources Committee voted recently to weaken protection for fishing in the Gulf.

The committee sponsored provisions that:
- Block progress in reducing the wasteful discard of red snapper and other important fish;
- Weaken protection for important fish habitat.

Call Senator Trent Lott (202/224-6253), who sits on the Senate committee overseeing America's fisheries, and ask him to work with us to fight similar weakening amendments in the Senate.

Otherwise, the Phrase "Gone Fishin'" Might Become "Fishing Is Gone."

Marine Fish
Conservation
Network
(301) 953-9111

acknowledged, was also a member of the Texas-based sport-fishing group.

In the *Sun Herald*'s report on the protest, its publisher defended the newspaper's editorial stance: "We certainly respect their right to get their message out, not only through pickets but through the media, which is why we published—and continue to publish—their letters to the editor.

"Our long-term concern for the environment led us to take the editorial stand we did, which called for a compromise that would serve the interests of both commercial and sport fishermen."

Compromise?

REALITY CHECK: TROUT OVERFISHED

At the December 19, 1994, meeting of the Commission on Marine Resources, in Biloxi, DMR biologist Tom Van Devender had told the panel that trout landings in recent years had been better than ever.

Apparently, they were too good—by the following spring the trout were overfished.

Data from the state agency's recreational creel surveys and fishery-independent sampling indicated that both the population and average size of speckled trout were diminishing. Department of Marine Resources deputy director Fred Deegen informed the commission that trout were "clearly overfished" and that some reduction was in order "to avert a crisis."

To that end, the commission, in the spring of 1995, proposed a round of new restrictions. When aired at public hearings, the cuts were generally agreeable to recreational fishermen with the condition that commercial catches also be reduced. But when the commission finalized the package at its September 1995 meeting, only the sportsmen took a hit.

"We've just been slapped in the face," said GCCA's executive director Scott Simpson. "We gave a lot, and nothing happened."

The CMR reduced the creel limit from 25 to 15 trout and eliminated a provision that let sport fishermen keep five fish between 12 and 14 inches;

These spotted seatrout, destined for the market, weighed an average of two pounds each because they were caught in a gillnet with 3 ½-inch webbing.

now all keepers had to be at least 14 inches long.

The commission also approved a 40,000-pound commercial quota. This was based on the netters' average annual harvest over the previous five years. Since the sportsmen's restrictions were projected to reduce their harvest by 30 percent, anglers wanted a comparable reduction in the commercial take, to 25,000 pounds. DMR biologists had in fact recommended a reduction in the commercial quota but more as a gesture to placate the anglers than as a conservation measure.

In 1994, recreational anglers accounted for about 85 percent of the total trout harvest. The count didn't include the substantial number of "regulatory discards," the undersized fish that died after being released.

The market fishermen's landings accounted for 15 percent of the total trout harvest. Moreover, the commission had just prohibited net fishermen from working on weekends, holidays, and near shore where the trout and most other fish were found. They had also been denied access to nearly all the barrier islands, which were favored trout haunts to which recreational fishermen retained year-round access.

Finally, the new 40,000-pound quota—which was being harvested by netters as well as rod-and-reel fishermen—represented a 45 percent reduction from the 73,000 pounds they'd taken in 1994, before the quota was instituted. A further reduction of 15,000 pounds might make the anglers happy but it would be inconsequential in rebuilding the trout population.

Prompted by its newly elected chairman, Vernon Asper, the commission stuck with the 40,000-pound commercial quota. It was a good starting point for effective regulation, said Asper, and could be adjusted as required.

Anglers in 1994 had taken over 405,000 pounds of trout. Managers, who didn't share the sportsmen's obsession with the commercial catch, had cut back on the sector that had the greatest impact on the population. In time, the reduced recreational bag limits and larger minimum sizes were expected to show results.

STANDING TOUGH

The new trout restrictions were to go into effect on November 1. A couple weeks earlier, sportsmen had made it clear that they'd had enough of the commission's environmental representative.

At a press conference in Gulfport, GCCA's chairman, yacht broker Keith King, announced that the "1,100-member" group was petitioning Governor Kirk Fordice to force Asper's resignation.

King complained that Asper sided with commercial fishing interests, as exhibited by his unwillingness to reduce their trout quota.

Though he'd been appointed to represent environmentalists, Asper had consistently voted against "environmental" positions, said King.

Asper didn't back down. He responded that he respected the GCCA and its position but said, "They need to do what they believe is right, and I will do what I believe is right."

Fishermen supported the scientist. "Dr. Asper is trying to protect the resource so that everyone in Mississippi can share in it, and he tries to be fair to all interests—sports, commercials, and the consumer," said SASI's Williams. GCCA was targeting him because he based his decisions on scientific data rather than emotion, and such adherence to truth was anathema to the group's tactics, she said.

"There is not a more dedicated environmentalist in the state than Dr. Vernon Asper, but he won't bend over backward for the GCCA," added Hilton Floyd Jr., communications chairman for SASI. "They can't control him, so they want to get rid of him."

The commission's netting regulations, which had taken effect in April, had severely crippled net fishermen, yet GCCA's membership wasn't satisfied, said Floyd. "Their obvious objective is to take all the resource for themselves," and they would use any tactic short of physical abuse to pressure commissioners who didn't agree with them, he said.

Indeed, GCCA was soon threatening the entire commission with extinction if the panel didn't give in to a long list of its demands.

Fishermen would spend the 1996 legislative session in Jackson trying to convince lawmakers to simply let the coastal commission do its job. Fallout from an untimely incident on the water didn't make their job any easier.

WILD WEST

Deer Island is the nearest island to Mississippi's coast and is more a partially submerged extension of the Biloxi peninsula than a real barrier island. Despite its proximity to the city, the undeveloped and wooded island attracted fish, particularly during the quiet of night when, in late January 1996, it attracted some net fishermen as well.

At 3 a.m., two patrolling marine enforcement officers stopped a Pascagoula netter, just inshore of the island. After citing the fisherman for safety violations, the agents heard another boat approaching in the darkness. The boat stopped a few hundred yards away. When the lawmen flashed their spotlight toward it, they observed that the boat had pulled up to a gillnet that was "illegally set," presumably because it was nearer to shore than the quarter-mile minimum distance that was permitted at night.

The boat's operator immediately cranked up his outboard and fled; the agents pursued the vessel until, they said, one of its occupants fired two shots in their direction.

"The pursuit was terminated because the officers were dealing with an unknown number of armed suspects at night," said Department of Wildlife, Fisheries and Parks enforcement Chief Terry Bakker, whose men then retrieved the abandoned net and notified the Harrison County sheriff's office.

The following morning, Ernest "Red" Ryan, of Helena, in Jackson County, reported that his boat was missing. "My boat was in Graveline Bayou in Gautier," he said. "I went there about 5 a.m. ... and the boat was gone."

Later that day, another fisherman found Ryan's boat on Jackson County's Belle Fontaine Beach. According to Ryan, his 24-foot boat had been stolen the previous night and stripped of the net, radio, antenna, tools, and other equipment.

Ryan denied involvement in the shooting episode and suggested that it had been fabricated. "I believe they staged the incident, that one of the enforcement officers fired the shots," said Ryan. "They said another officer checked their guns [to determine if they'd been fired], but that's like asking Frank James to check Jesse James's gun."

Ryan also complained that Bakker was illegally keeping his net while the case was being investigated.

"I would like to know how he knew we had his net or any net," Bakker responded. "There's a name on the net, but we're not saying whose name it is."

Bakker said that it wasn't unusual for someone who is suspected of fishing illegally to cover his tracks by reporting his boat stolen. "We recently chased a gillnet fisherman up the river. He stopped and fled on foot. He reported his boat was stolen 30 minutes later."

Lieutenant Bakker said he was asking outside investigators to take over the case. "We are going to get to the bottom of it."

No charges were ever brought in the case, but the dramatic affair was widely publicized. A *Sun Herald* article titled, "Gillnetters near Deer Island shoot at officers," said "Gillnetters threatened violence last year when the state CMR passed laws restricting their operation. Bakker said the Monday incident was the first evidence those threats could be carried out. Conservation officers have never been fired on.

"The officers had no doubt someone had shot at them. They saw the orange flash, then they heard the shot," Bakker told the *Sun Herald*. "They knew what it was. They're sure it was a commercial gillnetter."

While sportsmen made the most of the incident, SASI's Hilton Floyd tried to control the damage by reminding that no evidence had "yet been divulged and as yet no one has been charged." He also told the *Mississippi Press* that if Ryan's boat had been vandalized, it wouldn't be an isolated event. Commercial fishermen had been plagued with a series of vandalism, insults, and other incidents ever since the issue of banning gillnets was raised.

Blaming "a few radicals," Floyd related that Pascagoula seafood market owner Roy Martin's truck had been vandalized; gasoline had been poured on a $10,000 purse seine and set afire at Clark Seafood; Kevin Ryan's tires were slashed in Bay St. Louis; Pete Floyd's boat was stolen and set adrift; fishhook booby traps were rigged on crab pots of David Floyd; and a piece of steel was thrown at a commercial fisherman as he went past Ingalls Shipbuilding.

Floyd also told the paper that three commercial fishermen had been jumped and beaten by a gang of sport fishermen at Ocean Springs harbor in an incident that could have resulted in a major confrontation. But, said Floyd, "We are past the violence stage, and we are outnumbered several thousand to one."

MORE STRONG ARMING

The 1996 Legislature included a lengthy roster of fishing related bills: One would transfer marine law enforcement duties from the statewide Department of Wildlife, Fisheries and Parks to the coastal Bureau of Marine Resources. Another would totally ban gillnets from state waters by 1997.

A pair of identical bills—one in the House and one in the Senate—were offered as alternatives to the total ban. They each included a list of GCCA's demands that were simultaneously presented to the coastal commission; should the panel not approve those demands, it would be dissolved, with its regulatory power reverting to the Department of Wildlife, Fisheries and Parks.

Rep. Gene Saucier of Hattiesburg introduced the measure to totally ban gillnets. Sen. Tommy Moffatt, of Gautier, and Rep. Jim Simpson, of Long Beach, introduced GCCA's "compromise" bills into their respective sides of the Legislature.

The Mississippi Wildlife Federation joined the GCCA in backing the bills. The federation was the local affiliate of the Washington, D.C.-based

Hilton Floyd Jr., communications chairman of Save America's Seafood Industry.

National Wildlife Federation which had been a leading conservation group since the 1930s. The NWF usually tried to base its wildlife management positions on science but GCCA's siren song proved too alluring for the Mississippi chapter to resist.

The recreational groups' "compromise" bills included length and mesh-size restrictions on nets and pushed them at least half a mile offshore across the state. The bills also sought gamefish status for redfish, a ban on netting speckled trout after 1997, improved reporting, and limited entry for commercial fishing.

Moffatt had only recently been elected to represent Jackson County. When he admitted that his bill had been drafted by *Sun Herald* outdoor writer and charter-boat captain John Lambeth, who'd recently assumed the title of GCCA president, SASI's Williams unloaded.

"I say Sen. Moffatt has been taken in by the lies and propaganda of the silver-tongued devils. Lambeth is one of the slickest persons I know. I call it his poison pen. Lambeth's poison pen has struck again. He has duped our new senator," she told the *Mississippi Press*.

Moffatt defended his bill as one that would further conservation efforts while preserving the ability of commercial fishermen to make a living.

"Right now, we have too many people fishing too many nets in too many places," he said. "We don't even know what the landings are. This bill will have penalties for violating the laws. It allows commercial fishermen who have traditionally been commercial fishermen to keep fishing their nets. It gets the part-timers out," by requiring applicants to prove they earned at least 51 percent of their income from commercial fishing in order to get a gillnet license.

Such a means test penalized a person for trying to make a decent living, responded Williams, who said that Moffatt's bill, Saucier's net ban, and "several more bills" that were "coming out to further impede the commercial fishing business" had CCA written all over them. "And we wonder why in the world is the Legislature going to let this small group dictate fishing policy when the scientific data doesn't support any more fishing regulations?"

Such legislative efforts undermined the power and authority of the Commission on Marine Resources and the Department of Marine Resources, which were charged with setting restrictions, Williams said.

AFFLICTING THE COMFORTABLE

All journalists are familiar with the dictum that their job is to "comfort the afflicted and afflict the comfortable." CCA's legislative arm-twisting prompted a *Mississippi Press* reporter to do just that.

Rob Holbert had grown up in Gautier, gone to college at Spring Hill, in Mobile, and joined *The Mississippi Press*. After a stint as a reporter, he earned a master's degree in communications at Loyola University. As his "liver began to heal after leaving New Orleans," Holbert returned to the 24,000 circulation Pascagoula paper and resumed his coverage of city hall and crime. He also resumed his twice weekly column, "The Rostrum" which, according to his online biography, he'd named "after playing with the Thesaurus component of his word processing program for several hours."

Holbert's column was judged Best Commentary Column by the Mississippi Press Association in 1993. On February 11, 1996, his piece in the *Mississippi Press*'s Sunday edition was titled, "CCA tactics are going way overboard."

"On the whole, I would rank lobbying groups and automobile commercials among the most annoying things in the world. Both can be loud and abrasive and you always wonder if they're really telling the truth.

"Certainly, one of the most annoying groups, as of late, has been the Coastal Conservation Association (CCA). They are currently locked in an epic battle to save this state's fishing resources for the future. At least that's what they say they're doing. More and more, it looks as if the CCA is simply hell-bent on hammering Mississippi's commercial fishermen into non-existence. (Except for the charter boat captains. They're OK.)

"For the past several years, the CCA ... has advocated banning or severely limiting gillnetting in state waters. Their complaint has been that greedy gillnetters are depleting the natural resources we all should enjoy.

"While there's nothing wrong with an honest effort to make sure there are enough fish for future generations, the way the CCA has set out to accomplish its goals makes me think they should change their name to the Coastal Coercion Association."

Holbert admitted that when the issue first surfaced, his knee-jerk reaction was that "We'd all be better off without gillnets violating our natural resources. And it didn't help that many commercial fishermen are not the warm and huggable types. ... At the first meetings to discuss a net ban, some gillnetters issued threats and scared the (bad word) out of a few CCA members. This somehow lent credence to the CCA arguments that gillnetters were lawless caveman-types who could not possibly be trusted with the resource.

"But, time and again, scientific evidence provided by the Department of Marine Resources (DMR) has shown that it is not necessary to completely ban nets. 'Who the heck is the DMR?' you might ask. Oh, they're just the state agency charged with monitoring and protecting the state's fish stocks, that's all. Why would they know more than a bunch of sports fishermen?

"CCA leaders have repeatedly denounced the findings of fishery scientists who don't agree with their net ban point of view. They've reportedly made life pretty hard for some of those scientists. One may have even been demoted because of CCA pressure.

"Because the scientific evidence does not support a net ban, it seemed obvious that *reasonable* regulation of gillnetters is the best plan going. Enforcement of those regulations is also extremely important. But it seems unnecessary and even un-American for the CCA to drive these guys out of business. Unfortunately, that's what they seem determined to do. It's enough to make one wonder if the CCA is really a conservation association or just a group of fish fanatics.

"Like any good lobbying group, the CCA has gotten extremely busy in politics. Members spend a lot of time in Jackson twisting the arms of legislators. In fact, a current bill aimed at setting further restrictions on gillnetters was even written by CCA President John Lambeth. Wet-behind-the-ears Jackson County Sen. Tommy Moffatt introduced the bill, but it had CCA fingerprints all over it. Moffatt, an honest man, freely admitted that Lambeth drafted the bill. He should have realized submitting the CCA's bill is like letting the Chicken Producers of America write legislation regulating the American Beef Council."

The CCA had long hidden behind the "conservationist" tag, wrote Holbert, "portraying itself as a middle-of-the-road group that just wants the best for everybody involved. The real goal appears to be a total ban on gillnets and probably shrimp nets, too. I guess we're all supposed to eat seafood stamped 'Made in Mexico' so the CCA can win its war.

"If you think the CCA isn't interested in driving the commercial industry out of business, just read their goofy bill. If passed, it would make it almost impossible to work as a gillnetter," wrote Holbert.

"I'm sure many CCA members aren't really even aware of all the things their leaders are doing. But the CCA gang is pretty good at making it look like the whole world wants what they want. They're writing legislation, firing off angry letters to newspapers and otherwise pushing their agenda. Lambeth, a former newspaper reporter, even writes a weekly Outdoor piece for *The Sun-Herald*. (There's some journalistic objectivity for you. Guess what his favorite topic is?)

"Since the CCA doesn't have science on their side, they've decided to use money and political clout to achieve their means ... . You have to wonder what's in it for them. Somewhere along the line, the CCA moved from lobbying to zealotry, and that's probably what makes them so darned annoying."

LIMITED ENTRY

Limited entry, a qualification process that restricts the number of participants in an industry, was the centerpiece of both Moffatt's Senate bill and Simpson's House bill.

In 1995, GCCA's legislators in Louisiana—Mississippi's big brother in fisheries—had shoved a version of limited entry down fishermen's throats that proved more "limited" than "entry."

CCA had also forced Alabama's net fishermen to adopt the system. In Alabama, however, the leading commercial organization was dominated by full-time netters who didn't strenuously object to professionalizing their industry. Mississippi's group reflected that state's fishery, which had more part-time than full-time fishermen.

The few SASI members who favored limited entry were, not surprisingly, full-timers like Eli Ross of Biloxi. "It's the part-time pirates that give the commercial fishermen a bad name," said Ross. "They're the ones operating in the back bays and up around the sport-fishing areas."

Unlike in Alabama, where disagreement over limited entry had divided the industry for a time, SASI's members quickly reached a consensus— they didn't want it. Once they got on the same page, they focused on the matter at hand—the defeat of GCCA's bills in Jackson.

POWER VS KNOWLEDGE

When pro-CCA legislators in Mississippi introduced the group's limited entry/gamefish bills in the Legislature, they simultaneously presented the coastal commission with a list of demands that mirrored those in their bills. They also introduced a bill that threatened to abolish or de-fund the commission.

Commissioners took CCA's threat to disband the coastal panel seriously. The group had fought its initial formation and was dedicated to proving that the commission was a failure.

Sportsmen wanted to see everything handled in "Jaxon," one of the commissioners explained, because "The legislators can never be as knowledgeable as we are, and they are therefore much easier to bluff and coerce. If the folks who understand the problem don't agree with you, just take it to the next higher level of authority because they'll be less knowledgeable. Eventually you'll get to a group that knows nothing but has ultimate power."

Behind the scenes, some legislators suggested to commissioners that the CCA was only bluffing—it had taken them years of work to bring the CMR to the coast and they weren't going to reverse that effort at the drop of a hat.

At the same time, other influential legislators reiterated to commissioners that their extortionate bill was gaining traction even as legislators were withdrawing their support for CCA's "compromise" bills: Simpson's died in the House in mid-February; Moffatt's duplicate bill was defeated in the Senate.

However, despite opposition by fishermen and some coastal environmentalists, HB 1401, which called for dissolving the Commission on Marine Resources on April 1, 1997, passed.

After the defeat of CCA's demand-laden gamefish/limited entry bills, association leaders urged their members to "lobby, beg and threaten CMR commissioners" to pass the same proposals that the Legislature had shot down. The commissioners were held hostage by 1401. Unless they implemented CCA's provisions at their March meeting, nothing would be done to keep the law from taking effect and the commission would be history.

At the Commission's February meeting, with only five members present, the panel buckled to the sportsmen's demands, by a 3-2 vote. The result, however, was considered non-binding. A final vote before a full commission was scheduled for March 26, 1996.

Before that vote, the commission needed to hold a public hearing on the proposals.

SNEAK ATTACK

Although fishermen were already fighting for access in Jackson and Biloxi, they were not expecting a far broader attack.

At its February 20, 1996, meeting, Jackson County's Board of Supervisors passed a resolution asking the Commission on Marine Resources to make its netting restrictions uniform across the Mississippi coast by including the waters of Jackson County.

The CMR's original rule, which pushed netters at least half a mile offshore during the daytime and one-quarter mile offshore at night, had applied to the state's central coastal waters but exempted Jackson County from the provision.

Though most of the state's netters resided in the county, none were at the meeting when a pair of sport fishermen, one an engineer at Ingalls Shipbuilding, made the request.

"They caught us not looking as they shot us in the foot," said Jean Williams, who brought a couple dozen fishermen to the supervisors' next meeting. State biologist Tom Van Devender was also on hand, to remind the board that there was no scientific basis for the measure.

"You jumped on the hysteria bandwagon," Williams told the supervisors as she asked them to rescind their action.

One supervisor made a motion to do so, but it failed for lack of a second. Sport and commercial fishermen promised to show up in force at the board's March 20 meeting.

SPORTS AND ENVIROS

Meanwhile, nearly two hundred sport and commercial fishermen attended a tense public hearing in Biloxi where the adversaries traded taunts and expletives. The hearing was to air the slate of additional restrictions that CCA's leaders were pressing the commission to impose on the seafood producers.

The Mississippi Restaurant Association, Hancock County Seafood Producers Association, and SASI naturally opposed the proposed restrictions, as did individual speakers.

"This is an allocation grab, not a conservation measure," SASI's Hilton

Floyd told the commissioners. "It's like taking land from a family farm."

The Coastal Conservation Association and Charter Boat Captains Association endorsed their package of commercial restrictions while a local leader of a San Francisco-based environmental group told the commissioners that the sport fishermen's proposals didn't go far enough.

"The Sierra Club feels the compromise of a year ago is a failure because of widespread problems with enforcement," said Steve Shepard, chairman of the Mississippi chapter.

Shepard, a Gautier artist, was also a sport fisherman who sat on CCA's board of directors. "We want to see a total ban for a year and then allow it only if the Sierra Club can be involved in writing enforceable regulations," he said.

At minimum, Shepard wanted a 24-hour prohibition on netting within a half-mile of shore. He said the commission's nighttime relaxation of the buffer to a quarter mile put netters too close to the mouths of the bayous and marshlands, adding that the enforcement division of the Department of Wildlife, Fisheries and Parks considered the no-netting zones easier to enforce if they were the same 24 hours a day. The Sierra Club also wanted purse seines banned.

The Legislature had easily banned the seines for the harvest of redfish at the height of that controversy, in 1986, when their use was also prohibited within a mile of the shorelines of Hancock and Harrison Counties. In 1994, the Legislature disallowed purse seines for the harvest of mullet during the roe season, but they could still be used to take mullet outside of the spawning season as well as black drum and other food or bait fish including menhaden.

The local Sierra Club's official anti-commercial position reflected the consensus of most of its leaders. Indeed, two hardline executive committee members had resigned in protest the year before because other leaders within the club had helped write the commission's compromise that allowed netting to continue. They'd preferred an outright ban instead because there was "so much lawlessness" in the fishery.

Negative sentiment within the club wasn't unanimous—Asper, chairman of the coastal commission, was a member, and another of the club's greener members wanted it to stay out of the fish fight completely. A public statement of the group's position would have a negligible effect on the issue's outcome, she said, while it would certainly strain relations with the seafood producers who were working with her to curb pollution and other long-term threats to the fisheries.

"DON'T LET SPECIAL INTERESTS DECIDE FISHING"

Prior to the CMR's March 26 meeting, the *Mississippi Press*'s Holbert weighed in again on the issue. In his Sunday, March 10, column, "Don't let special interests decide fishing," he asked, "Wouldn't it be ironic if someday soon, consumers along the Mississippi Gulf Coast are munching on foreign-caught seafood because our own fertile waters are off-limits to commercial fishing?"

The local CCA was hardly a "loose-knit group of good ol' boys who got together one day and said 'Say, Jed, didja ever notice we ain't got as many fish as we used to? I bet them dang netters are the reason.'" The group was in fact in every state that touches salt water, "Except, probably, Utah," wrote Holbert.

"No doubt, the vast majority of those in the CCA really believe they are fighting the good fight to preserve fish species. Heck, all they want is to be able to take junior out and catch a few fish. But I think darker motives are what really push the CCA and its brethren across the nation."Seafood importers stood to make good money if the nets went "bye-bye," wrote Holbert. So did manufacturers of sport-fishing boats and tackle. "It only stood to reason that if sports fishermen were allotted the whole fishing pie, there would be more fish for them to catch and more people might become interested in the sport." The American Sportfishing Association, of Alexandria, Virginia, had even bought a lobbyist for the Mississippi CCA: "It's nice of them to take an interest in our fishery, isn't it?"

The ASA's president Mike Hayden had recently said that sportfishing was responsible for thousands of jobs in Mississippi and had an economic impact in the millions of dollars. Holbert concluded Hayden believed the recs were entitled to the entire allocation of fish and wrote, "Doesn't sound much like conservation, does it?" He also charged the CCA with using its "political cattle prod" to force the Commission on Marine Resources to consider another round of restrictions on netting.

"The CMR should do what it was designed to do," Holbert concluded, "which is regulate the fishery based on scientific data, not political arm-twisting. Unfortunately, the politically powerful and greedy forces behind the CCA aren't going to let that happen. They won't be content until we're all eating Mexican fish, while sitting on the edge of one of the nation's most fertile fishing grounds."

PROTECT RESOURCES OR BUTTS?

At its March 26, 1996, meeting, the Commission on Marine Resources was free to amend, adopt, revise, or reject Ordinance 5.013, which had been drafted by the state's Department of Marine Resources at the prodding of the CCA and with additional input from the enforcement division of the Department of Wildlife, Fisheries and Parks.

The ordinance contained so many restrictions that it was subdivided into three parts, according to the species that would be most affected. While those pertaining to the popular redfish and trout were rifled in intent, the proposals that were lumped into the "mullet" category were broader in reach and included the following:

Nets would be limited to a maximum length of 1,200 feet, a limit that was already in effect for individual nets; since mullet fishermen could legally connect two such nets, that practice would be prohibited. At the same time, the ordinance increased their mesh size. Under current law, the minimum was 3½ inches during the roe season, and 3 inches during the rest of the year. The new minimum for mullet nets would be 3¾ inches, and 4¾ inches for other species.

For the first time, a mullet season would be established; it would run just 45 consecutive days "beginning no earlier than October 1" and ending "no later than January 31 of the following year." Netting on weekends and holidays would be permitted during that season but would remain unlawful at any other time. First-ever commercial quotas for black drum, sheepshead, flounder, and Florida pompano were also proposed.

At the request of the Department of Wildlife, Fisheries and Parks' enforcement division, the no-netting buffer zones off Hancock and Harrison Counties would be extended statewide. "We felt that it wasn't fair for only two counties and not have it equally through the state," Sgt. Dean Lapoma told the panel. Although that coast-wide no-netting zone would be reduced to a uniform quarter mile offshore, nighttime fishing would be banned.

"I'd say over three-quarters of the arrests in regards to gillnets happen at night," explained Lapoma. "You just don't know what it's like to get out in the middle of an area that's closed, hit a Q-beam light, and it looks like a bunch of cockroaches running all over the place with illegal gillnetters in a certain closed area."

The enforcement division also requested the insertion of a *prima facie* clause. "We have got to have something with teeth in it to enforce to catch people," said the sergeant, who cited the example of a shrimper found five miles up a closed bay with a freshly retrieved trawl, muddy boards and jumping shrimp. Enforcement had no case, said Lapoma, because the violator wasn't caught with his net in the water, even though it was obvious what he was doing. Under *prima facie*, if it appears obvious, you're busted.

Set-nets were also targeted: to be considered "attended," nets would have to be directly attached to the fisherman's vessel instead of within a boat length.

The prohibition on netting within a mile of the federal Gulf Islands would be extended to Cat Island, the last of the state's barrier islands where netting was permitted.

The ordinance also contained a plan for limiting entry into the commercial fishery. In anticipation of Florida's net-ban vote, the 1994 Mississippi Legislature had passed its sort of Catch-22 law which prohibited issuing licenses to nonresidents from states that didn't issue the same licenses to residents of Mississippi.

Concerned that the law might be challenged in court, the coastal commission later passed its own emergency regulation that limited issuance of gill-net licenses to "individuals, firms or corporations that purchased Mississippi gill- and trammel-net licenses during any license year between May 1, 1990, and April 30, 1995."

Unless that regulation was revised to allow new entrants to the fishery, it would eventually lead to the demise of the industry. To preempt that process, CCA now offered its own plan:

To qualify for commercial net and hook-and-line licenses, individuals would need to have purchased such a license in one of the years 1993, '94 or '95. For license year 1996, they'd also have to verify that they'd derived at least 30 percent of their income from commercial fishing. In 1997 that would increase to 40 percent, and in 1998 and every year thereafter 51 percent.

According to SASI's Jean Williams, CCA's plan had the appearance of a compromise that would enable some gillnetters to remain in the fishery, but when combined with the group's other proposals, which included seasonal closures for all but 145 days of the year, it would in fact lead to the complete elimination of net fishing. "Who is going to be able to fish a third of the year to make half of their annual income?" she asked. "The income restrictions, combined with the season closures, can only mean one thing: by 1999, nobody will qualify for a gill-net license."

In addition to its comprehensive "mullet-related" package of restrictions, CCA's ordinance would classify redfish as a gamefish and reduce the commercial trout quota from 40,000 to 25,000 pounds. Market fishermen would only be permitted to harvest trout with hook and line.

Before the panel voted on the proposals, Asper lectured his fellow commissioners and the audience.

The ongoing argument over who could catch the fish "had a tremendous negative impact on all of us, our finances, our time, our sanity, our personal well-being," he said. "The biggest negative impact, however, of the whole debate has been to distract us from the real problem, and that is habitat loss."

The Gulf Coast had grown tremendously over the past two decades, and there were new casinos, more people, more industry, and more traffic in the coastal environment than ever before. "We've been building bridges," Asper said. "We've been paving highways. We've been bulkheading beaches by the mile. We've been pumping in sand. We've been filling in marshes. We've been dredging, until we have a landscape which suits us ... but it's extremely stressful to the fish that depend on the wetlands for their nursery habitat."

Instead of "hurling insults" and "accusing each other of decimating fish stocks," CCA and SASI should be working together to enhance the fisheries by addressing such environmental problems, he said.

As for the proposed ordinance itself, "Voting for it means that we agree with the notion that gillnetters are to blame for the ills of the fisheries. Voting for it means that we don't care about the long-term health of the fisheries; we're only concerned about meeting the short-term demands of a small portion of the population."

The panel was caught between scientific honesty and political reality, said Asper, but if the Commission on Marine Resources rejected the truth and based its decisions on cajoling and threats, "Where does that put us when we have to make more important and far-reaching decisions in the future? If we want to have any credibility, we have no choice but to back our scientists, to accept what they're telling us … because when the time comes, we are going to need to have their recommendations on other issues.

"In short, and in summary, we have got to … stand tough and face whatever consequences may befall us. We need to address the real issues, not the popular ones. We need to defend and protect our resources and not our own behinds."

The commission voted on each of CCA's three measures. They all failed to pass, with Asper, the chairman, each time casting the tie-breaking vote.

MIXED FEELINGS

Members of Save America's Seafood Industry had vowed to sue if they were subjected to any additional regulations. Instead, they sighed in relief. "It's high time the people managing the resource took the courage to stand up for the facts and stopped bowing to political pressure. This is the first time we've been before the commission that we didn't lose something," said Pete Floyd, a spokesman for SASI.

Sportsmen, on the other hand, seethed at their "defeat," and made a beeline for the Legislature.

"I think what we were talking about was a reasonable compromise," said the coastal commission's new charter-boat representative, Mikel Gusa, of Gulfport, who'd replaced Henry Boardman. Instead, Asper's actions had only made the situation worse and ruined any future chance of compromise, he said. "Phone calls are going up to the Legislature as we speak."

Rome Emmons, president of the CCA's Pine Belt Chapter, and scion of a Hattiesburg dry cleaning business, also considered his group's proposals to be reasonable attempts at compromise, after previously supporting a total ban on gillnets.

"We went to the Legislature, and they asked us to let CMR have one more chance," said Emmons. "I'm sure we will be going back to the Legislature. A total ban is an option."

In fact, the Commission on Marine Resources had already enacted a virtual ban but it would take another year for that to become apparent.

• CHAPTER THREE •

BIODEGRADABLE NETS

When Vernon Asper urged his fellow commissioners to reject CCA's wish list of commercial restrictions, at the panel's March 26, 1996, meeting, he reminded them that they'd already resolved the issue the year before when they passed their own package of netting restrictions. Most of those restrictions reduced the times and locations where fishermen could work, and took effect with little delay. Others hadn't yet had their full impact, he said.

"Nobody is talking about the fact that that ordinance contains provisions that outlaw non-biodegradable nets this coming January. That means after January the gillnetters are not going to be out there with monofilament. They're going to be using cotton or something similar. That's going to have a tremendous impact."

In its original wording, the CMR's 1995 ordinance mandated that gill- and trammel-net webbing be constructed of approved "biodegradable" material. As Asper had explained at the time, the intent was to "allow a certain amount of inefficiency into the system" by returning fishermen to the "olden days" when nets were made of cotton or linen.

In those pre-synthetic days, when fishermen returned from a trip, they hauled their natural-fibered nets from their boats—hand over hand—washed them clean of fish slime, limed them to kill bacteria and other rot-inducing agents, roped them onto scaffolds or reels and when they'd dried, loaded them back onto their boats, went fishing and then did it all over again.

A new gillnet with webbing made of braided nylon twine before coating with "Texasphaltic" net dip.

Immigrant and first-generation American fishermen in the "olden" days of the 19th and early 20th centuries typically worked from dawn to dusk and had plenty of help from their large families. They had another major advantage over Mississippi's 21st century netters—they could at least coat their webbing with tar or other preservatives.

The commission's ordinance prohibited net fishermen from using *any* treatments that might "increase the longevity of the net."

DN-103

When the commission passed its biodegradable provision in early 1995 fishermen weren't overly concerned—the delayed implementation allowed them to keep on working while, they hoped, their leaders would find a way to delay it even further. Moreover, they were holding an ace or thought they were.

BIODEGRADABLE NETS

Freshly dipped net. The tar-like coating protected the nylon webbing from abrasion and deterred rotting in nets made of natural fibers. Mississippi prohibited such coatings on cotton and linen nets, which made them even less practical for modern use.

The Shakespeare Company of Columbia, South Carolina, made polymers and fibers for a variety of industrial applications and was a major manufacturer of monofilament fishing lines.

Shakespeare claimed that it could produce monofilaments that degraded in seawater over any period desired by varying the concentrations of an ester in the plastic line. The company's DN-103 was designed to lose 50 percent of its strength over a year's soak.

There were some roadblocks to the fishing industry's adoption of the material. First, no net manufacturers were using it to produce webbing. Second, it "degraded" chemically and therefore didn't conform to the commission's definition of allowable materials, which were required to "biodegrade" through the action of living organisms like bacteria or fungi.

119

There was another hitch: DN-103 would create an enforcement nightmare because it was identical in appearance to the non-degradable and transparent monofilament that everyone was using.

Commissioners and agency staff wrangled over DN-103 through much of 1996 to resolve the hangups.

At its August meeting, the panel considered a one-year delay in implementing the ordinance's degradable amendment, to allow time for manufacturers to commence production of nets with DN-103, and for Shakespeare to develop a field test to identify the material.

The proposal required yet another public hearing in Biloxi. This one was lightly attended: eight people spoke in favor of the delay, and two against. Written comments included a petition with 216 signatures in favor of the postponement and three letters against it.

NO DELAY

In a September 13 letter, CCA state treasurer Pete Umbdenstock, who owned a Gulfport auto repair shop, castigated the executive director of Department of Marine Resources' over the proposed delay: "For your staff to recommend a one-year moratorium on the requirement for biodegradable gillnets should be an embarrassment to you and the commission."

If fishermen had been ordering nets that met the law for the past several months, "we wouldn't be in this predicament now," wrote Umbdenstock, who insisted that such nets were indeed available. "I have documentation of that. As a matter of fact, I suspect I have more time involved in researching this subject than any of the netters. What do they want, someone to buy the nets for them?

"I believe to further delay implementation will be one more example that someone isn't up to their job. I'm also sure the Legislature will be most interested in this chain of events."

Perhaps as a result, the commission decided not to postpone the amendment's effective date. The panel did agree to broaden its definition of acceptable materials beyond those that were "biodegradable."

In directing staff to craft that definition, Asper cautioned, "One thing we need to consider … is what the implications are going to be down the road. If we're requiring material like this for this type of net, it's not inconceivable that we could be challenged on this and also required to require this material for other nets and for recreational gear. So we need to consider what we're doing here, consider that this is a precedent-setting decision that we're making."

Commissioner Doug Horn, the Pascagoula fish dealer, urged haste so fishermen could gear up for the mullet run. "How are they going to be able to order something when they don't know what to order, and it takes them months to hang a net?

"If we can resolve it at the next meeting," said Horn, "let's do."

At its October meeting, the commissioners certified both linen and cotton as acceptable netting materials. The agency's executive director also reported on the public hearing that had aired the panel's proposal to allow the use of degradable synthetic fibers like DN-103.

The public comments included a letter from a Biloxi attorney who'd recently been retained by SASI, who wrote that since acceptable nets weren't available, the ordinance amounted to a complete ban on net fishing, which would cause an economic hardship for the state's fishermen. Even if nets of cotton or linen should become available, they would have to be replaced too often to be economically viable. And in keeping with the tenets of equitable fishery management, if degradable gear was to be required for commercial fishermen, it should also be required for recreational fishermen.

COMMISSION VOTES

At the panel's November meeting, Horn called attention to a recent event in New Orleans where restaurateurs were feeling the brunt of the Bayou State's recent net ban.

"Just day before yesterday, they had a meeting by the chefs in Louisiana saying that they don't have enough local fish, they're having to buy foreign and they're losing trade. That's the same thing that's going to happen to us," cautioned Horn, who felt that requiring nets to be degradable unfairly singled out commercial fishermen.

"We keep talking about these couple hundred gillnetters, but what about Budweiser and Miller Lite with these [plastic] cartons that they throw overboard? What about the recreational fisherman with his line when it breaks? He don't go back and get it.

"I think we are just going back to the same thing that we've been doing for years and trying to put these people out of business," said Horn, to a burst of applause from the audience.

Horn and his brother Phil had taken over Clark Seafood from their father Ralph; the company had been in the family for three generations. In addition to buying, processing, and marketing the catch of fishermen from around the country and south of the border, the company operated its own fishing vessels. Clark had purse seined bull redfish offshore until 1986 when the Legislature disallowed the high-volume gear for the harvest of reds. The Horns then shifted their efforts to other quality food fish such as black drum and mullet.

The Legislature eventually banned the gear for the harvest of roe mullet, as well. Black drum and a few other species could still be seined, "but it's not feasible to do that anymore," said Horn, because the cutting houses had "gone away" because there wasn't enough production to keep them open.

BIODEGRADABLE NETS

Biodegradable? Recreational fishing gear on high wire near a popular roadside fishing hole includes monofilament line, barbed hooks, spark plugs, Styrofoam floats, and plenty of lead.

Squeezed by such state-level management, Clark Seafood increasingly relied on federally managed fisheries. The company converted its purse seiners into trawlers to harvest deep-water butterfish and operated a fleet of hook-and-line vessels that targeted red snapper and other reef fish.

When the panel voted to approve the new degradability definition and adopt the ordinance with that language, Horn was the lone dissenter: "I just don't see how I can support anything like that unless it's for everybody."

DEAD END

From the get-go, commercial fishermen considered the commission's degradable mandate to be unfair and unnecessary, a "solution in search of a problem."

"Nobody from the finfishing side ever was in agreement with this crap," said Pascagoula fisherman Hilton Floyd. "At the time we were just ecstatic that they didn't get their ban, so we went along with the degradable thing, just because we had to. We were just optimistic that something would come through."

It didn't: When January 1, 1997, arrived fishermen had nothing to work with.

"My clients have checked throughout the country and to their knowledge no one makes such a net. You really have a case where it's impossible to comply," SASI's attorney Robert Harenski told a *Sun Herald* reporter in early January.

An unnamed net manufacturer told the same reporter that degradable gillnets would be expensive and difficult to produce. "It might be six months if we got on it today," he said. "What they have successfully done is stop commercial fishing in Mississippi."

After the ordinance took effect, Harenski filed for a temporary restraining order on SASI's behalf. A state court hearing on his motion was set for January 17. In the meantime, fishermen wouldn't test the law by trying to violate it, he said. "I have instructed my clients to follow the law to the letter."

By mandating degradable gear, coastal commissioners hoped to create demand for a product that businesses would then supply. But in business uncertainty trumps all.

In a January 9 communication, Shakespeare informed fishermen that DN-103 monofilament was "in prototype production" and could be "scaled up to commercial size samples and production quantities within eight to 10 weeks after receipt of product requests." Such requests required net makers to order a minimum of 1,000 pounds of line per diameter.

Owing to the gear's selectivity, fishermen needed an assortment of nets, each woven with line of a different diameter. "I have ten or so nets for the different fish I catch," said Pascagoula netter Devere Carter. "It's just like giving someone a single wrench and telling them to undo all the bolts on a car. You can't do it. We have to have different size nets."

Even so, Mississippi's fishery comprised a small market, and net making machines were huge.

"If you were a net maker and had a handful of net fishermen in Mississippi call you and ask you to redo your machines for a little run of nets and only a one-time order for the time being, would you invest the tens of thousands of dollars that it would take to make these nets?" Carter asked. "It's all business. The net makers don't even want to touch the DN-103 nets because the issue is in litigation right now."

In the meantime, fishermen tried to obtain webbing made of the untreated organic materials the commission allowed. In a January 15 response to them, the owner of The Fish Net Company, in Jonesville, Louisiana, said that with DN-103 on the horizon, no one was going to stock up on organic netting because they'd be stuck with a lot of unmarketable inventory.

"Therefore, I and everyone else in the net business are just waiting for the new materials to become available," he told fishermen. "I know that degradable is now the law in Mississippi saltwater gillnets, but until the new materials are available, I'm afraid you are out of luck."

Fishermen located a trove of leftover linen inventory at an Ocean Springs net company, and some other webbing that had been manufactured for purposes other than fishing.

"That stuff made of cotton is untreated and just falls apart. The linen stuff is like basketball netting. None of it really is any good for net fishing," Pete Floyd told *Mississippi Press* reporter Raines Rushin, whose January 27 article, "Gillnetters see change, trying times," marked this transitional period in the fishery:

> "Because of this, no fishing is taking place and the gillnetters are feeling the economic pinch. For most of the approximately 40 full-time fishermen in Jackson County, net fishing is something passed down by their fathers and grandfathers. Few know any other trades and some have already gone on the welfare rolls.
>
> "'A lot of fishermen around here are losing trucks and motors. I know of three guys who lost motors last week,' said Leon Martin, a lifelong fisherman. 'Ain't none of us never punched a clock. We're used to surviving. It's getting tough to do that though.'"

Roy Martin, owner of Martin's Seafood and Meat, told the *Mississippi Press* that he had to order mullet out of Mexico because none was coming in from the Gulf of Mexico.

"This is the only country in the world where they want to cut out folks who make a good living," he said.

The fishermen's plight met with little sympathy from the state or sport fishermen. "It has been up to the netters to make the orders [for the nets]," said John Henry, special counsel to the Department of Marine Resources. "Nobody will be sitting around making a net unless an order is placed for them. Cotton and linen were approved a long time ago, but evidently none of the netters were interested in using them."

"This was done in lieu of a net ban—it was put forward by the

Fisherman Kevin Ryan exhibits deteriorating linen trammel net, after just weeks of use. *(Courtesy of Hilton Floyd)*

commercial interests," added CCA's Pete Umbdenstock. "This was their deal and now they've chosen to sit on their hands and wait."

With the fishery at a standstill and the future uncertain, one fisherman told the *Mississippi Press* that he was leaving it in God's hands.

"The gillnetters try to make this a biblical thing—Jesus casting the nets into the water and all that," said Umbdenstock. "Jesus's nets weren't made of monofilament."

On January 29, the Nylon Net Company, in Memphis, informed fishermen that it had located a company that would weave webbing from 27-pound-test DN-103. Tests were being run and "We hope to have netting in six to eight weeks," the company said. "We will need booked orders to get material," at around $15 per pound.

Nearly a month later, Shakespeare advised the fishermen's attorney that the company hoped to have "trial samples" of 15-pound test DN-103 ready "sometime in early to mid-March 1997."

Then Mississippi's Department of Marine Resources took Shakespeare's high-tech monofilament off the table.

The commission required any webbing of a material other than cotton or linen to lose 50 percent of its tensile strength after one year's immersion in salt water. After an initial accelerated test, the DMR had found DN-103 acceptable. Later, at the conclusion of a real-time test, the agency found that the line had not lost 50 percent of its strength and declared it illegal.

"It's good-lookin' stuff and we were all thinking that was gonna come through, but it never happened," said Floyd. "It never happened."

Sportsmen later convinced their legislators to push gill- and trammel nets a half-mile off all of the state's shorelines, including Jackson County's. "We had that linen net crap going on so that didn't quite kill it," said Floyd. "But they came back and got that half mile off any shoreline through, and that was the last nail in the coffin."

SPORT FRENZY

While netters bled, sportsmen flooded the 1997 Legislature with another round of anti-commercial bills.

One, House Bill 1580, had nothing to do with netting but marked another loss for fishermen and seafood consumers.

The cobia was a popular restaurant item on the Gulf Coast, where it was usually listed as "lemonfish." Commonly four to five feet long, the shark-like fish were attracted to buoys and other floating objects where they were targeted by recreational and commercial rod-and-reel fishermen.

Charter-boat clients with cobia in 1964.
(Courtesy Maritime and Seafood Industry Museum)

The legislature had outlawed the sale or purchase of Mississippi-caught cobia in 1990. HB 1580 tightened the sportsmen's grip on the species by formally declaring it a gamefish.

Other bills in both the House and Senate called for stricter penalties for netting violations and expanding the areas where the gear was prohibited.

"The only way to make sure you have a say on the outcome of these bills is to call your local senator or representative," a *Mississippi Press* outdoor columnist advised his readers. "Most of these bills deal with gillnets. Like many years past, bills concerning gillnets have made their way to the Legislature but rarely is one passed that would do some good."

MISSISSIPPI FISHERFOLK
IN THEIR OWN WORDS

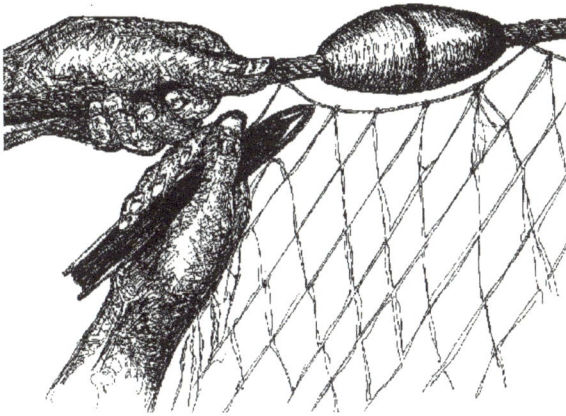

MISSISSIPPI FISHERFOLK
IN THEIR OWN WORDS

JEAN WILLIAMS
PASCAGOULA

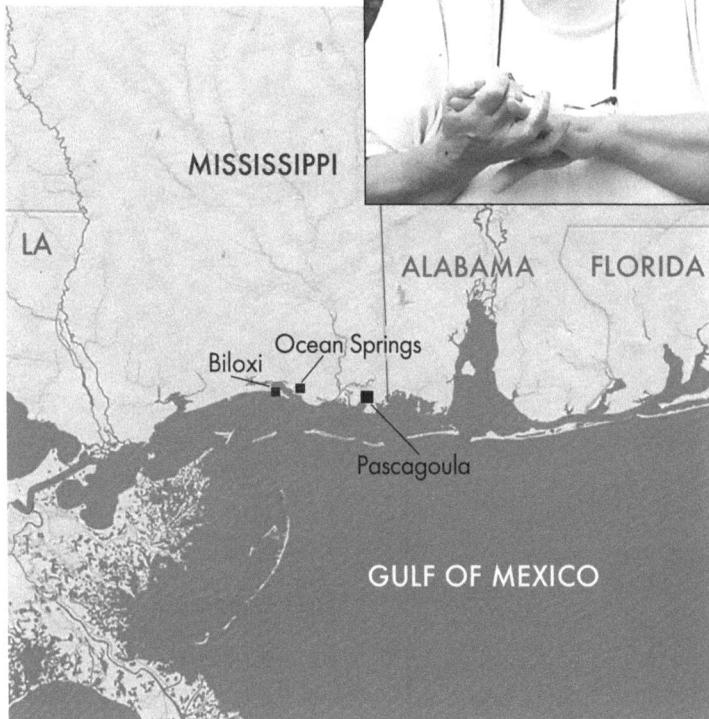

133

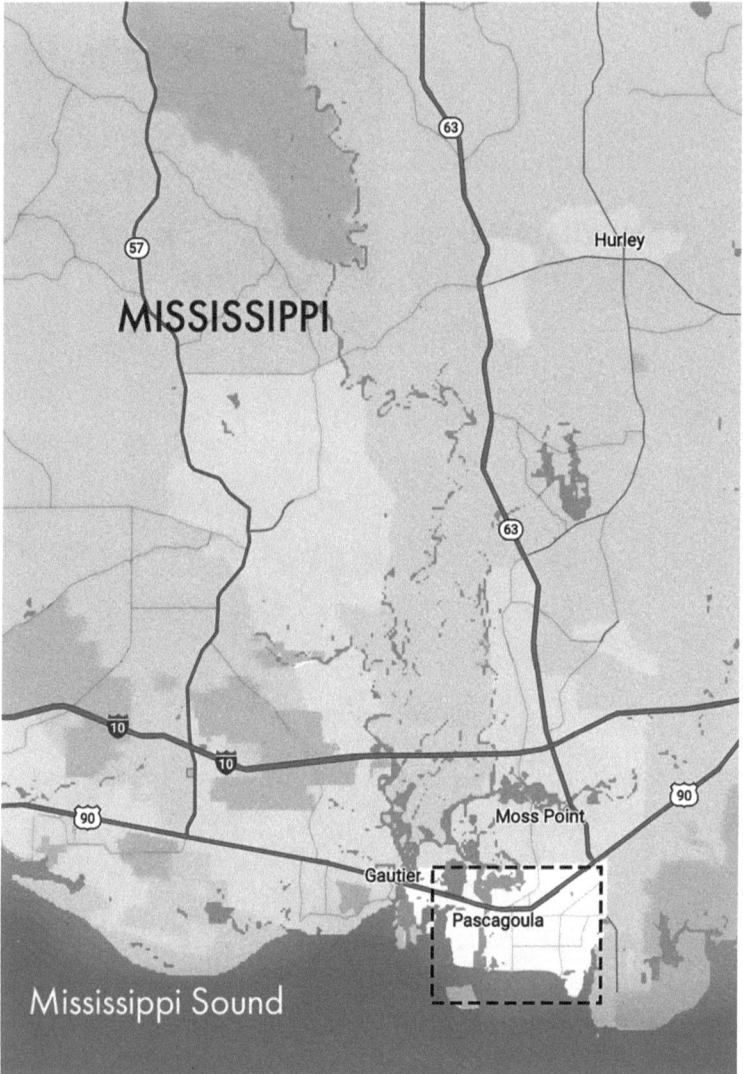

MISSISSIPPI

Hurley

Moss Point

Gautier

Pascagoula

Mississippi Sound

JEAN WILLIAMS PASCAGOULA

Jean Williams married into a Niceville, Florida-based commercial fishing family in the 1950s. As her husband, Clarence, followed the snapper and grouper, the couple relocated first to Morgan City, Louisiana, and then to Pascagoula, where they settled in to raise their children.

In Pascagoula, Jean's husband fished for the Horn family's Clark Seafood. Times were good as Williams worked the reef fish offshore, and even better during the mid-1980s when he hit the heyday of the bull red fishery aboard a purse seiner.

Federal regulators "interrupted" the harvest of redfish in federal waters in 1986, and a few years later instituted the first-ever management restrictions on the reef fishermen's staple—the red snapper. The income of the Williams family nose-dived. In defense, Jean, who'd previously contented herself with the duties of a traditional housewife, founded Save America's Seafood Industry to provide Mississippi's offshore fishermen with a say-so in the management of federal fisheries.

By the mid-1990s, the state's inshore net fishermen were flocking to her organization as they tried to fend off Mississippi's insatiable sportsmen.

The net ban battle was white hot when I first met Jean, at a book signing event she'd organized for me at the Singing River Mall in Gautier, just across the river from Pascagoula. The cab of her pickup bristled with radio antennas, its rear bumper was plastered with pro-seafood bumper stickers, and the front seat was stacked with letters, newspaper clippings and reports. As Jean stepped down from the truck, this unassuming lady, with glasses propped low on her nose, wore a white sweatshirt printed boldly with her group's acronym—SASI.

By August 2003, when I last saw Jean, she was 68, her organization had unraveled and, after a recent shoulder operation, she wasn't feeling too sassy. I didn't have the heart to ask her to pose in her sweatshirt, but we did spend an enjoyable afternoon watching an old fishing film that her husband had taken on his snapper boat back in the 1960s. Crammed with nautical mementos, fishing photos and artwork depicting commercial fishing scenes, her living room had the air of a cozy seafood museum, and it was there that she told me her story:

I grew up in West Virginia—in the middle of the state, toward the north, about 40 minutes from Morgantown. On the water? Yes, a river. A dam had been put there in the early '30s by the WPA. It was just a beautiful, great place to grow up.

One day a bunch of us started across the river. The guys in the boat went a little faster than I could, swimming. So, I got about halfway across, and I started screaming. Well, they realized I was in trouble and started to come back, but I gave out before they got to me.

I was about 12 or 13 and I said, "Dear Lord, don't let me drown." And I just let myself down and there I was on a big flat rock, in the middle of that river.

As soon as I graduated, I tried to get into the Air Force, but the Air Force had met their quota. So I went into the Army in about 1950. I got stationed at Fort Gray, North Carolina, and my good-looking little husband was at Fort Gray too.

He was in the Airborne and worked as an ambulance driver and a lifeguard. He was so dark from the sun that the only reason I knew he was a white guy, his bathing suit was down just a little bit.

My girlfriend and I had been swimming on a lake and to get back you either had to walk around through the woods or take a boat. So Clarence and his buddy said, "Yeah, we'll take you back." They rowed us around the lake and tried to get us to go out with them. But my girlfriend and I, well, you didn't just date the folks you had just met, especially guys in the service. If you were in the service you came with an automatic stigma back then. But they finally talked us into going out with them.

We wound up going out to a drive-in that night. He went to the concession stand and came back with his shirttail out, filled with everything he could possibly carry in it. "I forgot to ask you what you wanted," he said to me.

Was his dad a fisherman? Oh yeah. Clarence is a fourth- or fifth-generation fisherman. And when the first of our boys married, he told him, "If you want, we'll get you a boat and I'll help you pay for it until it's yours."

But Jimmy said, "No Daddy. No disrespect, you've provided for us great, but there have been too many times when we didn't have a daddy at the ball games and things. And I don't want to put my family through that." So that ended it.

After we got married, we went to Niceville and rented a little duplex thing—like a little summer camp. We lived there awhile and then, since his daddy was already working for Ralph Horn, they decided that Clarence should run a boat. So they bought the Bob, a little bitty snapper boat.

Clarence was fishing snapper out of Morgan City and the Horns would send trucks over from Pascagoula for the fish. When he'd come in, I'd drive over to be with him. I'd throw those two little boys in the car and follow him everywhere. Of course, that was before interstates. And one time I was going over there, and I was just barely into Louisiana when they got me for speeding. Well, while I'm there with the cops, who comes around but teenager Doug Horn, in the truck, blowin' that horn.

And, of course, there was no way I could get to the dock before he did. So I was in hot water when I got over there!

We moved to Morgan City for a while. Our third son was born there. And then we decided that it would be better to move over here to Pascagoula. I said, "I'll go on one condition, if you go ahead and buy a house."

He came on ahead of me—because I was expecting our third child—and picked out our house, down on Country Club Drive. So, when I came over, I didn't even know where I was going. I didn't know where Country Club Drive was, got turned around two or three times and finally stopped a lady and she took me there.

"Well, this must be it," she told me. "This 'Sold' sign." And so we raised the kids over on Country Club Drive, which was a good place. It was so isolated. Safe, you know?

We lived there about six or eight years and then we moved here. We've lived here for 40 years. How many kids? Six! Four boys and two girls.

And none of them has gone into the fishing business. But Jimmy has a marinated crab business. He has "Bayou Blends," here in Pascagoula, with a partner. They buy the meat and then they have a process. And it is scrumptious, I'm tellin' you. They can it in jars: "Bayou Blend Marinated Crab Claws."

It's the end of the claw, where you have to scrape, you know. If you send some to Yankees in Michigan, you have to give 'em instructions. Which we did, one year for Christmas—we sent everybody some and we got all these calls, "Well, how do we eat it?"

It's been a real struggle for him. He and his wife got into predicaments where they never lost anything, but it's come pretty close. Down to the wire, you know? About like fishin'.

They're doing well, they just hit these snags. And, of course, the down market is about to come. A bad time of the year. He's so close to being there, he's so close. I tease him, "Yeah, he's gonna be a millionaire one of these days." He's this close to being a millionaire. Or a pauper.

Back when we lived in Morgan City, Clarence had all kids on that boat with him—16-, 17-year-old boys. They'd stay out there and they'd work their behinds off if he would promise to get home for the weekend. They'd party all weekend and show up on Monday morning and go again.

But that's one of the things that they've always said about him, he's never been a "captain" captain. He's been a participant in all the work.

When I got involved politically, he was fishing with Jim Reahard on a purse seiner that Jim and the Horn's owned. Later on, they were dragging for butterfish when they hit that treasure. That was on the Mistake, the boat the Horns had built to purse seine redfish. When they closed that down, they converted it to a stern trawler and were using it to catch butterfish. Clarence was supposed to go with them on that trip but another fellow was running the boat and he made Clarence nervous. So he didn't go out. And here they hit this treasure with the trawl—a bunch of silver coins. Now that's the story of our lives.

They still have that little museum there in Grand Bay, Alabama, for the little treasure that they hit in 1993. [The El Cazador museum focuses on the history of the New Orleans-bound Spanish ship that sank in early 1784.]

Why did I get involved? I got involved because it was wrong. It was wrong. My first [Gulf of Mexico Fishery Management] Council meeting, I went to New Orleans. And I went to the podium and I told 'em that when our youngest daughter started tenth grade, my husband went out and bought her a brand-new little red Fiera. Now this was, what, four or five years later, I said, now she's in college and he can't afford to send her spending money. And that's the honest-to-god truth.

And then I broke down, then I lost it. And, what's his name? Corky Perret, the biologist with Mississippi DMR. And the guy out of Texas, with the shrimp fleet? Julius Collins. They were sitting there together and Julius asked me a couple of questions, and Corky said, "This is what this council needs to hear more of."

You know, me standing up there bawlin'. But since then, I've learned to control myself.

How did SASI get started? SASI was me. I called up Myrna Reahard and her daughter-in-law Cindy—there were five of us, I think. I called Kay and I told her we were going to get together and talk about it. Kay is married to Clarence's baby brother. And Kay called me back and she said, "No, I don't want to get involved. Bobbie really doesn't want me to do it." I said, "Kay, if you want to do this, it's time."

One of the things that the Williams men do is they suppress their women. Or at least they try. Poor Clarence hasn't done very well with his!

I said, "Okay, we'll meet at such and such." And she showed up. And the first couple of things we did, we went to a couple of hearings. We went to Alabama, and we really didn't know anything about it. But when we got over there, Kay said, "Well, I could put down a few things." And honey, she waltzed up to that microphone, and she was a beautiful girl, hair long down her back, and it worked. I mean all these men were just, well, it worked to our advantage.

SASI rocked on and rocked on, and we were doin' and doin'. We took in a few people and then we had a good group. But there were a couple of "shit stirrers," for lack of a better word.

We got a lawyer, we thought he was gonna do us some good so we started collecting money for that. And this one old gal got to questioning the money and, really and truly, all but said that I took their money.

You know how that goes.

The other part of it was, I was getting little grandchildren out the yin yang and they needed a grandmother. You know, I had family obligations. I resigned, we had an election and elected new officers. I told them, "I will help you any way I can." But the next few turns of events just stopped that altogether.

To begin with, Cindy was the treasurer and she's a good girl. She never touched any of the money and neither did I. But what happened was, they wanted the books. They decided that we had to turn the checkbook over right that minute. Cindy wanted to make sure that, well, she was a perfectionist, and she wanted to make sure that every "i" was dotted. Well, they got downright ugly about it. One thing led to another and by the time we got the transaction over with, I was sick to death of all of it. And so was poor little Cindy.

We did spend money without, you know, getting the gang together and asking every time. We'd ask, "Are we going to do this?" and I'd say, "Yes, write a check." But it was always accounted for.

Once we decided we needed a fax machine. Everybody had voted that we would get the fax machine. So I drove to Sam's and bought one. And it was about $500, I think. It was a whole lot more complicated than I needed but I brought it home. Jimmy hooked it up and I said, "Okay, show me how."

He said, "Oh, Momma, you don't need this thing. You just need a little one." So I packed it up, took it back to Sam's and got a cheaper one.

Well, the next thing I know, we show up at the meeting and somebody brought up that we needed to buy another fax machine because if we could afford to buy one for $500 then we could afford to buy two. You know, stuff like that, just crappy stuff.

Finally, things just went from bad to worse. And it just petered out. I don't think they've kept the post office box anymore. It all just sort of went away. I think maybe, had I remained a member ... but it was just taking its toll.

But I'm really proud of what we did. And it wasn't *me*, it's what *we* did, as a group. We came together and we fought 'em. And I think we put up a good fight but there's just no winning. And there is no way to bring it back.

Getting back to all of this, the federal Council system and all that, to me, that is the biggest farce that was ever invented. Because I feel that all they do, it's just smoke and mirrors. I learned that we didn't make any progress. We held 'em off for about five years but it's just too traumatic, too draining, to go up there and sit for a week and listen to all that. And then to think that you may have made a difference. And then go back to the next council meeting and they have wiped it all out.

And then they have the Council meetings in all the fancy hotels and what it becomes is, for lack of a better word, just a little boys' social club. And everybody gets caught up in it and everybody's friends and all this, and there's no way out of it. And I had people tell me this before we ever got started. But I thought, "No, surely, we could do something."

I wasn't actually on the council, but my sister-in-law Kay was. I just went to speak, and to stir the pot. Who was that professor over there at South Alabama, that was so involved with the Council? Bob Shipp. I made him so mad one day he threatened to throw me out. He threatened to throw me out if I opened my mouth one more time. And guess what I did? God gave me a big mouth, I reckon. And I use it too much.

One thing that we did here, we got rid of a sportswriter in Biloxi. We run that sucker slap dab out of town. He was horrible. I mean he was just so one-sided, just lies, lies, lies. So we took him on, and I was sending letters to the editor and one day I was fixing to roar off another one, and

I was talking to the lady at the paper who was in charge of the letters. And it got dead quiet, and she says to me, "Miss Williams, I think I ought to tell you that I'm his wife."

"Oh, you poor darlin'!"

We got up to Jackson and he jumped me after a meeting. We'd gone up before one of the committees up there, the Senate, I think, and after we got out, we were just socializing, shaking hands, and he said, "I'm not shaking your hand." One thing led to another, and he said, "You know what you've done? You've attacked me personally."

I said, "Wait a minute, how more personal can you get, where you've attacked my pocketbook? You know, *that's* personal."

If you'd have seen his face. I had some fellas surround me that were ready to get him if he made a move. I told 'em, "I appreciate that, but I can handle it."

He owned a charter boat and was writing the outdoor columns. So it ended up that we went over there and picketed 'em. Just had the biggest rally you ever saw, with everybody honking their horns and everything.

The whole deal was, the *Sun Herald* bought out this newspaper over there some years ago. The old man who owned the paper was the daddy of this lady that was married to the sportswriter. And so when he sold the paper to some corporation—I don't remember the name, the same one that's in Florida and all—the stipulation was that she would have her job. You know, she would still be part of it.

But we run him slap dab out. He went to South Florida. I don't know if she went with him or if she saw the light.

Then we had this other little guy, Rob Holbert. He worked for our paper down here, *The Mississippi Press*. He was going to all the meetings at the Coliseum and everywhere we'd go, the commission meetings.

They had him at the meetings, holding him by the arm, walking him around. They were wining and dining him and taking up all his time.

But I got right up in his face one day over there and I said, "Now look, I want you to know that what's goin' on, they're lyin' to you. They're not tellin' you the truth." And I told him. "All this is wrong, and I can prove it." And when we proved it to him, he came onboard, and he was great.

Before that, he was writing what they were telling him to write but doggoned if that kid didn't wind up writing half-page editorials for the fishermen and making sure that things that were said were right.

His daddy is a doctor, and I had worked with him, so I had a bit of a good rapport. But that's what I always did, what I always tried to do was just be truthful. I didn't have to backtrack lies because what I was talking about was really the truth. I researched it or I wouldn't say it.

I think Rob went on to go to Washington to write speeches for Trent Lott. I'm assuming that's where he is now.

[After a stint in Washington, D.C., as U.S. Sen. Trent Lott's deputy press secretary, Holbert returned to the Gulf Coast, settled in Alabama, and co-founded *Lagniappe*, a feisty weekly that grew into the largest locally owned publication in the Mobile area.]

We had a good thing goin' here because Hilton Floyd was a very good writer. And Hilton would say to the fellas, "Can I use your name?"

"Yeah. Put that," they'd say.

It got to where the fishermen would say, "Hey, Hilton, you got a letter for me?" It went really well, and we did do some good.

Did Chevron give the CCA a bunch of money? I don't know anything about that. But there were some guys at another company that got into a bit of a fix. They got into some trouble with the company because they were using stationery with the company's letterhead to promote the net ban.

We got two or three letters goin' about them supporting it, and why, and that they were doing it on company time. And we got that shut down. No, it wasn't Ingalls, it was a smaller company. But we got with

Ingalls some too. We made them take down the posters. Jimmy went with me to do that.

They were puttin' up posters, "Support the Gulf Coast Conservation Association" and all this stuff. Just the crap that shouldn't have been put up.

We got Wal-Mart too, because they had on their bait freezer out there, "Join the GCCA." I went in there and straightened the manager out. He's my buddy, you know?

Damage control. It was constant, you had to constantly do something.

The environmentalists? Yes, they tried to help us a little bit. We had worked with the Audubon Society and the Sierra Club, with Becky Gillette. They had a joint meeting, and it was on wastewater. And we went. People from all three counties went, and we had about 20 or 25 fishermen there. We had more people than the environmentalists had!

And what we went there to tell them was how terribly important the environment is to us. But that we had selfish reasons for wanting to keep the estuaries clean, because this is where our product grows. And what we told them was, "No, you don't see us out fighting pollution. Because we don't have time because we're too busy defending ourselves against GCCA."

We told them that we definitely wanted to work with them on the environment but that we have to spend all our time trying to put out fires, the lies and propaganda that's always being presented about us.

Pascagoula was a humongous fishing town. Even when we got here, both sides of the river were lined with shrimp boats. All kinds of fishing boats. It generated, oh I don't know, a lot of money. It was just really a lot of fishing families. And a lot of characters, you know? I guess a lot of 'em ended up going to the shipyards. You know how that goes—a steady paycheck.

What's Clarence doing? He's been working on a pogy boat for about ten years now. He fishes for Omega Protein, over in Louisiana.

I saw on TV the other day they had a tournament going on, for redfish. And they were out there in their boats, "There they are! There they are over there!" And they'd cast and they'd hook one, a big old red. They got four more big reds before they ever got one that was small enough that they could keep. And I thought, that's just play. What a waste!

It's gonna be fine for all these sportsmen, they can all go out there, they can catch fresh fish every day. But where are the little old ladies in the middle of the state going to get their fresh fish? And I feel like that right now—where am *I* going to get my fresh fish now?

CHAPTER FIVE

MISSISSIPPI FISHERFOLK
IN THEIR OWN WORDS

PETE FLOYD
PASCAGOULA

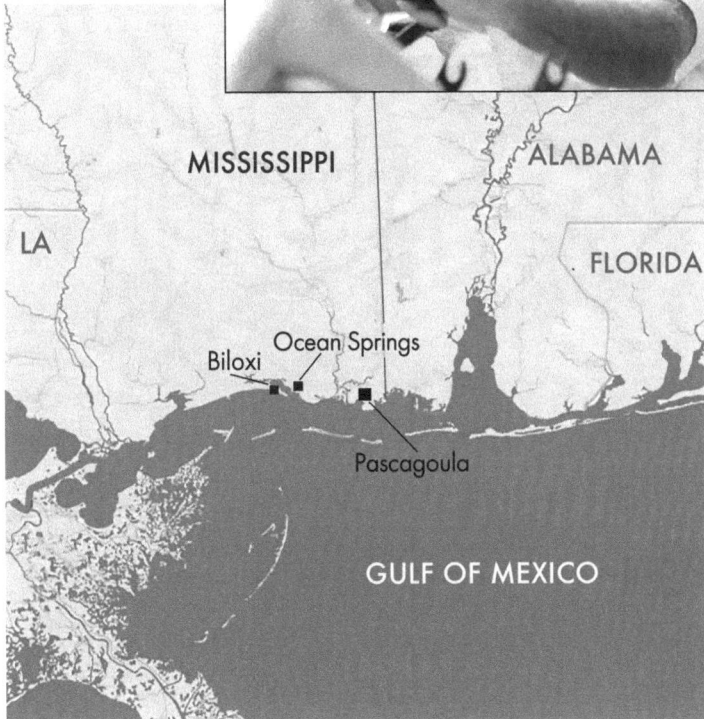

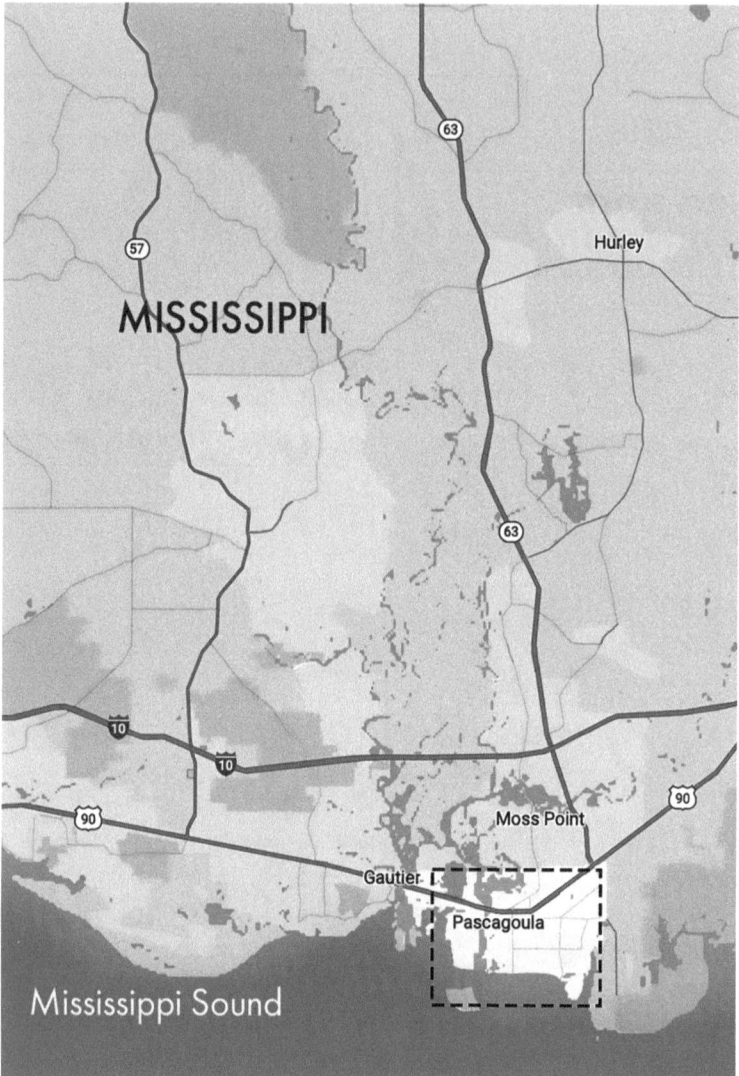

The battle over nets in Mississippi was fought within the Legislature in Jackson and the Commission on Marine Resources on the coast. The battle also raged on another front—the editorial pages of coastal and capital-city newspapers where correspondents attacked and counterattacked in their letters-to-the-editor. Many of the letters from the seafood industry's side were authored by either Pete Floyd or his younger brother Hilton.

The brothers attribute their writing abilities to their mother, an English teacher with a passion for the written word.

"Mom was a book lover," recalled Pete. "She started when she was real little, reading books. And she told me one time that when she was just a little bitty girl, she was so surprised when she learned that people wrote books. She said she thought they came from God."

Mrs. Floyd herself authored a few booklets about the old days in Mayport, Florida, where her husband had fished commercially. Captain Hilton Sr. later worked for the federal government, running exploratory fishing vessels in the Gulf and the Atlantic, and as a fishing methods and gear specialist.

Fishery biologist Harvey Bullis, who was chief of the Gulf Fisheries Exploration and Gear Research division in the U.S. Bureau of Commercial Fisheries, (which would become the National Marine Fisheries Service), first hired Captain Hilton. The two became friends and when their families got together for a picnic, Pete, who was 15, first laid eyes on Jeannie Bullis, who was 13. A few years later, the fishing and scientific families merged when the teenagers married.

The Floyd family traces its fishing lineage back many generations and, in their own ways, each of Pete and Jeannie Floyd's three children earned their livelihoods from fish.

Daughter Traci, their oldest, worked as a marine fisheries biologist with the Mississippi Department of Marine Resources, where she worked her way up the ranks to become director of the department's Office of Marine Fisheries in 2022.

Scott worked with the U.S. Fish and Wildlife Service where, among other things, he used hoop nets, trotlines, and gillnets in pursuit of Alabama sturgeon, which was endangered due to the damming of its riverine habitat.

David, their youngest, completed a few years of college but quit to go fish commercially full time. "He'd have a hard time up there in college in north Mississippi when the mullet would get to running down here," said Pete.

Pete's brother, Hilton Jr., was a commercial net fisherman. After the net ban, he kept on the water with a job as an engineer with a menhaden outfit. During the winter off-season, he cast-netted for mullet, sheepshead, and other inshore species.

Pete himself fished full time until the late 1970s, when an early wave of fishery restrictions prompted the young father to hire on with the local fire department.

On his days off, he fished. A self-taught naturalist, he also parlayed his childhood passion for catching snakes and turtles into the occasional contract with university museums and the Nature Conservancy. Even

(*Opposite*) Jeannie and Pete Floyd.

152

beachcombing, another of Pete's childhood passions, brought in a few dollars.

Pete and Jeannie reside in a shaded neighborhood in her family home, kitty-corner from former U.S. Representative Trent Lott's beachfront house. In August 2003, surrounded by museum cases packed with his collections of fossils, antique bottles and other beachcombing trophies, Pete talked about his fishy family, the good days, and the bad:

Our family came from Minorca, an island off the coast of Spain, in the 1700s. They came to the East Coast of Florida and were indentured servants for the Turnbull Plantation in St. Augustine. As years went by, they turned to fishing. So, my family goes back fishing, many generations.

Way back then, when they came from Minorca, they carried the wife's and the husband's name. So our name was like "Floye, hijo de so and so." And through the years over here it just became "Floyd."

All my uncles, even the uncles that married my aunts, were connected to the sea—net builders, fishermen, pilots, bar pilots. One of my great grandfathers was a bar pilot on a sailing ship out of Jacksonville.

We were living in Mayport when I was born. And I was raised beach-combin'. Literally, from the time I could walk, we were beachcombin'. We'd all go down to the beach and Papa and my uncles would be pulling beach seines. Actually, they did the seining on Atlantic Beach, Neptune Beach, Jacksonville Beach, toward St. Augustine. And they had what they called Mayport Cadillacs.

They'd take these old '40s and '50s cars and not long after they bought 'em, the trunk lid came out and they built a bed like a pickup truck on the back of it. With bin boards and all, just like a shrimp boat.

They'd back these "trucks" down to the beach and load 'em up with the fish after they'd pull in the seine. It was neat.

And while they were working, my mother and my aunts would take us little bitty kids to look for shark's teeth on the beach and look at sand peep eggs and stuff. You know, just real naturalists.

Papa was mainly shrimping back then but also beach seining, a little bit of gillnettin', cast-nettin'. And he got a job offer from the U.S. Fish and Wildlife Service to be a captain on one of their research vessels.

My wife's father was the guy that hired him—Harvey Bullis. He eventually retired as director of the Southeast Fisheries Science Center in Miami but back then he was director of the Pascagoula lab. When he first came to the house to hire Papa, I was six or seven years old. And Mayport had "horned toads" —Texas horned lizards. I think some of the fishermen were fishin' in Texas and they brought a box of 'em back and let 'em go in the sand dunes of Mayport and they took up there.

Anyway, I came in with some and gave 'em to Jeannie's dad, and that was the first connection with my wife. When he came back to Pascagoula, he gave Jeannie the horned toads that I had caught. She was two years younger. She may have been 5 and I may have been 7. And then I didn't actually meet her until I was about 15.

We moved from Mayport when I was nine, to Brunswick, Georgia, where the Silver Bay was tied up. Papa was on the Silver Bay, the Pelican, they even offered him the Oregon II, but he turned it down.

They strictly did exploratory fishing in the Gulf and the Atlantic, and Papa's full of stories. They discovered over 350 new species of sea life while he was captain of those boats. They fished with any kind of gear, but they drug trawls a lot and they pulled them in up to 500 fathoms of water. Every drag they pulled up was full of stuff nobody'd ever seen before. They discovered the royal reds, all sorts of stuff.

I caught my first sea turtle on the Silver Bay, with a handline. That's the only sea turtle I can say I ever killed. Well, I didn't actually kill him. I was just nine and I pulled him up alongside the boat and they lassoed him and pulled him up and gave him to somebody on the dock. And they just ate him. That was back when people still ate sea turtles.

The Silver Bay was tied up to the dock in the Brunswick River, I was fishing and that's where they come to nest. They nest on all those islands along the Georgia coast there, and quite a few come up the rivers to nest too. That turtle was way up that river, a loggerhead.

We stayed in Brunswick for two years, and when I was in the sixth grade, Papa got transferred to the lab here in Pascagoula.

That was when I pretty much started fishin', cast-nettin', flounder giggin', frog giggin', catchin' snakes, catchin' turtles. I even caught cats and sold 'em! Coons, 'possums.

When he got transferred to the lab here, Papa changed jobs. He was still with the government, but he got off the boat and became a fisheries methods and equipment specialist. He would write papers, like how to build a cast net, all sorts of stuff.

Then he went to Vietnam as a fishery advisor. He was in all those fishing villages, trying to help them improve their fishing methods. He ended up with an awesome slide collection, with pictures of all those Vietnamese fishing boats, with those eyes painted on 'em.

He was there at about the time the war was winding down—when they were moving into Saigon. He said he'd be in his hotel room and hear the bombs goin' off. I guess he was over there for about a year before he had his heart attack.

After that, he got a nice retirement. His retirement was based on what he was making over there, and he was pretty high up on the GS level. But we were gillnettin' and he gillnetted a lot with us. He couldn't make much because he was retired on a disability but he would go with us and he would set us up and help us.

He used to love fishin'. He still does but he's gettin' too old. He's 79. Just moved to a new house, he got him a dock on the water and mullet jumpin'. We were just checkin' his crab pots for him. Oh, he's in heaven.

Papa was a heck of a navigator. He knew how to read the bottom, not just from his job on the research boats but as a shrimper. Back then, they had those old Raytheon depth recorders and they watched 'em like a hawk. They learned to read the bottom with those little chart recorders.

Like I say, I was in about sixth grade when we moved here, and we'd never fished here. So, we get the gillnets and stuff and Papa gets out a chart. And he looks at the chart and he says, "This is where you need to fish." Never fished here in his life, looked at this chart, and when they banned gillnets, 30 or 40 years later, I was still fishin' where he'd put his finger on those charts!

I did kind of the same thing later on. We went to the [Florida] Keys, and I wanted to go lobsterin'. I got a chart and I got to lookin' at that chart and it said, "rocky bottom," close off the shore from where we were going to stay. Well, I got a rowboat, a little eight-footer and rowed out to where it said "RKY," and I found these big old brain-coral heads and got our limit of lobster.

Believe it or not, nobody was there in lobster season in the Keys. You could see the boats runnin' everywhere offshore, stoppin' and puttin' the divin' flag up. But everybody missed that one little spot of coral.

We gillnetted around here, then we moved back to St. Simons [Island, Georgia], after Papa was transferred back over there. This was at the beginning of twelfth grade, and I stayed over there about four months and then I ran away from home and came back to Mississippi. I went back to gillnettin'. And I did a lot of shrimpin' back then, out of little skiffs. Anything, really, to make a living.

During that time, Jeannie and I got married. I was in twelfth grade and she was in tenth. And then it was time for me to go to college, so we moved back to Brunswick. I went to Brunswick Junior College and worked half a day with the University of Georgia's Marine Extension Service.

I worked for Dave Harrington. He was in charge of the University of Georgia Marine Extension, and we worked together every day.

I worked for Dave in 1971 and '72. He was just a fantastic guy, and a real good friend. But in '73, he had a cancer in his stomach the size of a grapefruit and they gave him six months to live. I was at Skidaway when they found the cancer, and I came down to visit him in the hospital. I fainted when I walked in, the only time I fainted in my life.

He was lying in the bed and I didn't think it was him and when it hit me I went to my knees. After I came to, he started laughin': "Some friend! Come to cheer me up and faint." But he beat the odds—he went through the chemo and lived another 30 years.

I made several trips on the Georgia Bulldog. We trawled for rock shrimp and did a little exploratory fishing. We also went exploratory fishing for swordfish, but we used a swordfish boat for that. It was called the Gulf Stream, out of Portland, Maine. It was an old side trawler that they converted to a longliner and brought down to see if there were any swordfish off the Georgia coast. We caught 20 swordfish and 500 giant sharks, and we were just fishing with about six miles of line.

The size of those sharks! I cut the jaws out of one and I could put 'em all the way over my body without touching. Somebody came by the house and had a fit over 'em, and I gave 'em away. I wish I had 'em now. I can't remember if they called that one a dusky or a silky. It was a big old offshore shark. We caught a lot of blues, and one mako. All we caught was his head—the other sharks'll eat a mako.

I made two trips on the research boats, and you talk about fun. One trip I just tagged along with Papa. And the other one was for the University of Georgia. I was taking specimens for their collection—

anything that NOAA didn't want, I would get for their museum.

We drug trawls a lot in real deep water. And when they'd dump them out on deck, the only thing you'd recognize would be Coke cans—marine litter—because you'd never seen any of those fish or crabs or shells, or any of that before. You'd see things that resembled the inshore stuff, but were very different, like an offshore sea robin. Then you've got your black swallowers, viperfish, all these strange looking things.

And where the bottom got too rough to pull a net, they'd drag shell dredges—no suction, just a big oyster dredge, maybe eight feet wide, with teeth. And they'd bring up a big pile of fossil shells, mastodon teeth. Oh, it was neat. You'd have a huge pile on deck, and it was *all* fossils. The most gorgeous sharks' teeth you've ever seen. They were sort of copper colored from being out in the ocean for so long.

What happened to that? Oh, they went through it, it's in jars and stuff all over the country, I imagine. I'm sure the Pascagoula lab still has a lot of it. A lot of it went to the Smithsonian. Papa discovered *a lot* of stuff out there. They discovered a shell, if I remember right, its scientific name was *Pleurotomaria*. It was thought to have been extinct for like 300 million years. And when they dumped that shell dredge and those *Pleurotomaria* dropped out, still alive, the biologists who were there and knew, their eyes popped out. They had a fit. It was like bringin' up a coelacanth.

I worked out of Brunswick for two years and then I went to Skidaway for about a year. When they decided to make the University of Georgia Sea Aquarium, there on Skidaway Island, they wanted me out there to do that. I really didn't want to leave Dave and them, but Dave talked me into it. He said it was a once-in-a lifetime opportunity and he was right.

We actually set the tanks, built the filters, the whole works, and then went out and caught the stuff. We had an ichthyologist, a Harvard teacher, you know, a big shot. He was the curator, the big guy at the

COASTAL EROSION PART ONE

aquarium, and he would tell me where to go, to catch fish for the tanks. Now here's an ichthyologist that, I mean, seriously, this guy didn't know shit about fish.

Oh, he'd tell you how many fin rays a fish had, or how to key a fish out, but when he'd tell me to go to this area to catch this kind of fish, heck, you wouldn't catch nothin'. So I'd leave and I wouldn't tell him. I'd go catch the fish that we needed and I'd come back and say, "Yeah, man, I got 'em right where you told me."

But that's how a lot of the biologists are. Really and truly. And that's one of the problems with fishery management today. A fisheries biologist has a degree on paper and that's all he's got. I have yet to meet very many that had a clue as to what was going on with the fish. And I've known *a lot*. Being raised with the Fish and Wildlife Service, you know, I've been through it all my life.

There are a lot of things they know—on fish structure, morphological things, or something like that, they blow me out of the water. But tell one of 'em to go catch an Atlantic sharpnose and you might as well tell him to build a spaceship.

And that's why you end up with these damned regulations. Because they say, "Okay, we're going to go out and see how many redfish are out here." Well, they bring somebody, a potato farmer from Iowa, not to insult anybody but, realistically, that's what winds up out there, in many cases, doing the research.

And we see them, you know? They come and put their nets in places that a commercial fisherman would be *insane* to try to make a living catching a redfish there. The wrong time of year, the wrong place, everything's wrong, but they put these nets out, or traps or whatever, and do a six-month survey: "Oh, these fish are in dire straits. They're on the verge of extinction!"

Well, no shit! Where you're fishing, they've never been there.

You couldn't catch one if your life depended upon it.

Like right now, I don't know how it is in Lou'siana but July and August, go try to catch a redfish around here. You ain't gonna do it. In the Mississippi Sound, they're just not here. They're either up in the marsh, or out at sea. I think most of them are out at sea right now. Well, I know most of 'em are out at sea—they're catching them out there: "Stringers," four pounders or so. Not the real little ones, the 12-inchers— they're up in the marsh. But the larger fish, there are giant schools of them, with 50,000, 100,000 pounds to a school. But they migrate and, unfortunately, there isn't enough data for the biologists to know that.

I've had the biologists that did the research on the redfish, the ones that did the flyovers in the Gulf of Mexico when they banned it in federal waters? They have told me that there was nothing scientific about it. That the data is so flawed it's unreal. Because they had an agenda.

And it's not their fault, they honestly believe what they're writing and what they're fighting for. And they'll argue with you about it because what does a dumb-ass commercial fisherman know?

The latest one that really got me going, I saw a paper that said the Atlantic sharpnosed shark was in terrible shape because of the pogy boats. Bycatch. And at the time this paper came out, we literally couldn't fish for the Atlantic sharpnosed shark. They were frenzied in the whole [Mississippi] Sound, everywhere. And here you've got these shark experts saying that there are *no* Atlantic sharpnose. That they're in trouble because they couldn't catch 'em. And here you couldn't put a damned hook in the water without catching an Atlantic sharpnosed shark. Literally, swarms of 'em. It was the right time of year for that, they do it every year, in certain areas.

It's not to sound egotistical, but we always caught fish because we had to catch fish. And we stayed on the fish. You know, if I quit fishing a certain fish for two weeks, I've got to figure out where he is all over again.

That's why a lot of sport fishermen are unsuccessful—they can't fish as much as we do. They go out twice a month and if they don't catch any fish, we're the cause. They see us come in with a thousand pounds of fish in the boat and it kills 'em. Or they ride by and see us pickin' up our net and we're pickin' fish out of it, and they're catchin' nothin' because they don't know what they're doin', and they've got to have somebody to blame it on.

It's kind of like we do the same thing. We didn't catch any fish because the water was too clear, we didn't catch any fish because it was too rough, everybody's got an excuse. Well, their excuse was us!

"They're catchin' 'em all, they're wipin' 'em out." But the tiny percentage of the fish that we were takin' was nothin'. Nothin'!

One of the things we dug up in that respect, during the net ban, we dug up some data from Florida that indicated, well, proved that more speckled trout were killed in Florida through catch and release than were landed annually by commercial fishermen. The Florida Department of Natural Resources is where I got it from—they did a study on it. Aside from the fact that the recreational fishermen were landing 80, 90 percent of the fish to the dock, they were killing more than the commercials were catchin', through catch and release, because a lot of times, the fish goes into shock and dies.

If you throw a fish back, he swims off and looks great. But what they did, they penned some fish up and the next day they were all dead. Or most of them were dead, because it stresses 'em so badly, they can't make it.

If you notice these fishing shows, if you look at the fish they throw back, a lot of times they don't show them swimming off. Or they'll drop 'em in the water and you'll see that fish in shock, going sideways to the bottom. Especially with billfish and stuff like that, which just becomes shark bait.

Yeah, you've got to follow the fish, stay on them. Fish aren't going to be in the same place. That's like these people that think they're gonna go to this wreck and if they go to that wreck and they don't come in with a bunch of big fish, some commercial fisherman has their fish. The fish didn't use his fins and his tail to swim to a better spot that happened to have a better salinity, or more food, or to spawn or whatever.

"No! A commercial fisherman got 'em. That's why I didn't catch any. Hell, I've got a $50,000 yacht that I can sit up on the flying bridge and drink martinis, and you're telling me that I can't catch a snapper? Well, hell, it's got to be somebody's fault. I'm rich, you know? I can do anything I want!"

I used to have them come by me when I was netting mackerel outside of Petit Bois Island. After we got the aquarium all set up on Skidaway, we moved back here to Pascagoula. And that's when I really started fishin.' Gillnetting Spanish mackerel was my main source of income.

They're like maggots out there but these big old yachts would come by me, trolling some kind of goddanged jig that a damned boobie wouldn't bite and they'd cuss me. Here I am in a little bitty Bosarge skiff with a 25-horsepower motor with a thousand pounds o' mackerel in the boat. He's on this big yacht with a bunch of bikinied women, about six rods and reels trollin' for Spanish mackerel and he's not catchin' anything.

And there would be literally so many mackerel that if they could have seen what was swimming underneath their boats, they would not have believed it. But what they were trolling behind the boat at that time of the year, a mackerel wasn't going to bite. And I used to hear 'em cuss me, "Goddamned gillnetters got 'em all!"

Like that idiot in Alabama wrote years ago when I was gillnettin' out there. He used to write for the Mobile paper, and he wrote this article— I can remember to this day the way he worded it. Now this was in the late '70s when he wrote, "It seems like every time a sport fisherman starts catchin' a few specks, the commercial boys get their binoculars out and come rope us in with gillnets."

I was catchin' Spanish mackerel! Fishin' way off the beach where you couldn't catch a speckled trout if your life depended on it. But he wrote this huge article, and he was talkin' about me, you know, he described my net and everything. Boy, I was so mad.

I fished the islands year-round until the late 1970s. At least three months out of the year, I mackerel fished. I had a real good market for mackerel fillets. There was a restaurant here in town that was buying them from me. I sold them as many as 900 pounds in a day. They'd use the fillets for their buffet. And I was getting 95 cents to $1.00 a pound for them which was just unreal, back then. And easy fishin'. I knew right where the fish came through and when they came through, so I'd go and run my net out and usually pick it right straight back up. Then all I had to do was fillet the fish.

I had a campsite out there where I'd stay. I was by myself, had the whole damned island to myself. I did a lot of night setnetting out there. Mainly whiting and trout is what I was focusing on, right on the beach.

A lot of the times if it was a pretty night—in winter when the mosquitoes weren't there—I'd just sleep out under the stars in my sleeping bag. And one time, for some reason in the middle of the night, I just opened my eyes and a damned raccoon was staring me right in the face. Scared the crap out of me, because he was right there.

Then they said you couldn't fish out there from May 15 to September 15. That was the original thing. That would have been the federal people, probably the people with the national seashore here. I'm sure a lot of sport-driven politics is what it was. That was pretty much before the net ban thing got kicked off hot and heavy. That was when it was just startin' to get a little scary.

164

When they shut down the islands, things really got tough. That's when I went to work at the fire station. See, back then, you couldn't sell sheephead. Mackerel were hard to sell. You pretty much had to focus on redfish and trout, flounder and whiting—popular stuff like that—unless you lucked up and found little markets.

Besides that restaurant, I had the ship chandler over in Mobile who was buying a lot of the fish that we couldn't normally market. He was selling sheephead to the Greek ships to make soup and stuff out of.

After they shut down the islands in the summer, the state shut us down on the weekends and the holidays. You know, no gillnettin' on weekends or holidays. Then, no gillnettin' within 1,500 feet of anybody's pier. Then you had to be tending your net all the time. In other words, I couldn't have gone up on the beach and gone to sleep while my net was out.

They just came up with one thing after the other. I can't even remember all the regulations that were just hittin' us, hittin' us, hittin' us, on the way to the net ban. Taking a little piece at a time, restricting redfish, restricting flounders, then they started coming up with all the size limits. It was just like a snowball coming down a mountain. It got bigger and bigger and bigger.

The next thing you know, the sport fishermen started to come out of the woodwork. I guess the economy got good, and the islands, on weekends you couldn't have fished out there on weekends anyway, the way I was fishin'. They'd have tore your nets to pieces.

We already had the sports just a runnin' through the nets. They had these big high-powered boats, with 150s and stuff, and they used to take great joy in running through our nets and cutting the corkline. Purposely. They knew we couldn't catch 'em in our little boats.

I used to beg 'em, I'd wait for 'em to come back. They never would though. One time I had four of 'em run over my nets. I begged 'em to come back. They would have killed me if they would have, but I'd have put a couple of 'em in the hospital!

Next thing we know the net ban really kicked off, they shut the islands down permanently, and things started *really* getting tough.

Yeah, the net ban was pretty doggoned nasty. You really see what kind of people you're up against. I heard of some vandalism over in Lou'siana but there were just some little minor incidents here, like my brother Hilton's truck gettin' keyed, you know, people eggin' David's truck here at the house. Little stuff like that. I found big piles of nails under my tires. It's *still* goin' on. They just egged David's truck a few months ago. I know who did it, without a doubt—these CCA officers' kids in the neighborhood, the Hitler Youth.

The ones that I know don't care about their kids, they go off fishin' with all their buddies or to the yacht clubs and drink, and the kids are left at home to do whatever. A lot of 'em are hooked on drugs and just in trouble all the time. Fightin', you know, just problem children. Their dads are too busy bein' great, warped people, givin' 'em everything, you know?

It's a big difference from us. We were raised workin', having to work for what we've got, and they were handed everything. And when they hop in the boat and they see this poor fisherman out there catchin' more fish than them, that ain't right. "I'm wantin' *his* birthday present, I don't care if it is his birthday, I want *his* present."

What the CCA did, they convinced the sport fishermen that the fish are their fish. And if you've read their literature, I'm sure you've seen that: "They're takin' *your* fish. Don't let these netters take *your* fish."

And there's no compassion for anybody. They don't care. Like these commissioners here, when I'd get up to talk to these commissioners at the hearings, and in one-on-one conversations with these guys, I'd tell 'em, "You're killin' families that are out there that have never done anything but fish."

166

They don't have any education, they quit school before they graduated from high school, but they know more about fish than any biologist in the country. They know their jobs, they learned it through generations from their daddy and their granddaddy. And when you take this from them, they're destroyed. These families are completely destroyed, and I've seen it firsthand.

I mean it's horrible what's happened to some of these people. I'm more fortunate because I'm into snakes and turtles and research. I've done so many different things that I could go in different routes, to make up for the fishin' that I'm missin'. But these poor people, they're just shocked. You ought to see some of my fishin' buddies. They're just ruined, completely ruined, for no reason whatsoever. It's sad, it's just really sad.

They had all these hearings, which were mostly for show because it was all decided before. Over a thousand people showed up at this one, mostly sports. CCA. They were all dressed in their little sport-fishing uniforms, wearing sunglasses around their necks, at night, the double-brimmed hats, Ban-the-Nets T-shirts, the whole works.

And there was this little girl, about 12 or 13 years old. Her daddy's Martin Young, the third or fourth, from an old fishing family. This little girl had sat down and written a little thing she had wanted to say, about her daddy fishing. She gets up and starts talking, "Don't take my daddy's job, let my daddy fish, we need the money," and all this stuff. And she was real nervous, and kind of shaky, and she started cryin'. And when she started cryin', all the big CCA people started yellin', "Poor little baby. Where's my violin?" You know, just no compassion for this little girl, whatsoever.

A lot of people jumped up and it was fixin' to get nasty. But they had police in there and, fortunately for them, we're nonviolent people. Of course, there were times when I've wanted to rip their heads off. I mean, verbally, I got pretty nasty with some of 'em. If they'd have twitched their eye, I probably would have killed 'em.

They were always telling us that we could be guides. We could be fishing guides, we could take the people that have slandered our families for decades and take 'em in our boats and show 'em where we caught our fish. Like we would let one of them step in our boat after they called us "environmental rapists" and everything else, "a pirate taking booty from the sea."

It was amazing how widespread that sort of crap was in the '90s. What blew my mind, I was watching Teenage Mutant Ninja Turtles, which I never watch. I just happened to see it at the fire station for some reason. And here the turtles are after this shrimper, he's a pirate takin' booty from the sea.

I mean the whole damned nation, any media they could hit was against us. In little subliminal things. Even the movies. Hollywood. What was that movie, with that commercial fisherman? The guy beat his wife all the time, he abused his daughter and all. Of course, they used a fisherman for that. The timin' was perfect, right when they were puttin' us all out of business, portray the fisherman like a villain.

CNN? I don't watch it much anymore but back during the net ban I used to watch it. It was full of stuff against the commercial fishermen. CNN slammed us relentlessly.

Around here, you won't see many CCA stickers at launching ramps. There are not many CCA people here in Pascagoula. Now where they have their little headquarters is by the casinos and stuff, where they're all protected. Where they've got all the big yachts, you'll see them. But they go to these hearings and tell our lawmakers that they represent all the sport fishermen.

Well, it's pure bullshit. See, they have these fishin' rodeos and if you buy a ticket for the fishin' rodeo you're automatically a member of the CCA. So they can go to a commission meeting and say "Well, we have 150 members." They don't have 150 members, they have 150 people that entered that fishin' rodeo. They gave 'em a sticker, they stuck it on the truck and they drove off and they never saw them again.

The CCA, the head guys, overall, are very rich. They're the company owners. They get everybody in the company, when they come out with a petition or something, they get everybody in their company to sign it: "You better sign it or you're gonna be in trouble with the boss."

Republicans? Not necessarily, at least not here in South Mississippi. The CCA, when you get over in Ocean Springs, which is really the headquarters, they are extreme environmentalists. Democrats. At least that's my idea, that more liberals are environmentalists. Plus, you see, they cross party lines. That's what really hurt us—you had Democrats *and* Republicans workin' together against us.

They're just people that don't have a clue, people that don't care. Like Schwarzkopf. That's how Florida lost. Well, that's one way. Of course, they took that Georgia Bulldog footage, and they showed the footage of the dead dolphins in the driftnets in the Pacific. So people thought, "Oh shit, these gillnetters are killin' all our porpoises and sea turtles!"

I never, honestly, killed a sea turtle in my life in a gillnet. And that is the truth. Almost 40 years of gillnettin', never killed a single sea turtle. Now I've seen a lot of them come up in shrimp trawls. But I've never seen a dead one come up in a shrimp trawl.

What happens with the sea turtles, they get where you're draggin' and they've got all this bycatch. They're gonna stay there eatin' all those little fish. I've found them where they were completely chock full to the throat of little croakers and such.

What used to happen, before the TEDs [Turtle Excluder Devices], you'd get one shrimper that catches him, and the next shrimper catches him, and by the time the fourth shrimper catches him, well, I suspect that's when they go. But then again, every dead sea turtle that washes up on the beach is blamed on the shrimper. Well, how many of them got tar balls? How many got plastic in their stomach? How many got hooks in their stomach?

That's something else we dug up. In Monroe County, in the [Florida] Keys, the extension service there did a study and found that 75 percent of the sea turtles killed in the Keys were killed by sport fishermen. You're not going to see that on CNN.

Do I have a degree in herpetology? No, it's all self-taught. My mother bought me a reptile and amphibian field guide when I was about five years old, because I liked to catch turtles out in the ditch. By the time I was nine I was pretty good at catchin' snakes and identifyin' snakes.

I just had an interest in wildlife which started on the boat, looking at the piles of stuff that came up in my daddy's trawl.

They loved the birds, you know, they loved the dolphins, they showed me all the things about them when I was comin' up. When there'd be certain birds that came around the dock, we'd name them. We'd get to know individual dolphins. We'd be out there fishin', there'd be one with a nick out of his fin or somethin', and he'd come to the net. Every time you'd set in a certain area, he'd be there, and you'd toss him a fish.

Commercial fishermen are all naturalists, regardless of how they're being portrayed now. They're the mountain men, the mountain men of this century. And last century. The last mountain men. Because everything we do is in the wild.

When we're out there fishin', we see birds, we see stars, we see dolphins teachin' their little ones how to fish. And we like it. And we love these animals, literally love these animals. They're our best friends, they're our neighbors.

And then these damned CCA people come portray us as people that hate them, kill 'em for fun like it's a pleasure hunt. When in fact, they're the ones that are out there torturin' and killin' these fish for fun. It's just beyond belief how successful they were at portrayin' us that way and turnin' it around, when they're the pot, well, they're not the pot callin' the kettle black, they're the devil.

And I watch 'em on these fishin' shows when they're out there catchin' these fish and gigglin' like a bunch of little kids. They're holdin' a big old trout or somethin', that's in shock, they fought him for 45 minutes on a flyrod with six-pound test where he's *gonna* die, and they're all "eee, eee, eee," slappin' each other on the back. Well, they don't have any compassion for that fish. They don't even look at the fish. They don't even look at the fish's eyes.

When we were kids, when we'd get in trouble around the table or somethin', a thing in our family was, my daddy would tell the other brothers and sisters to give him the fisheye. There were seven of us and they would all look at you with a sad look, like a dyin' fish. You just rolled your eyes down like a fish layin' on the bottom of the boat.

That was our punishment. There was not much spankin' goin' on, a fisheye was worse.

But the point I'm trying to make is, commercial fishermen have compassion. We don't like seein' a fish layin' there dyin', it's what we do for our living. Where these guys like watchin' a fish die, and watchin' a fish suffer at the end of a line. That's what they live for.

It's going to be interesting, what it all comes to, what we're gonna do. It's gonna change. In one of these catalogs I order fishing stuff from, I read that they were hangin' people in the 1400s or 1600s, for pullin' nets. If you were caught seining in Spain, or one of these European countries, they would hang you. There was a Ban-the-Nets movement centuries ago. Sure was. History repeats itself.

Yeah, this is pretty emotional for me, and I get wound up. During the net ban, it got to the point where we couldn't sleep at night. Jeannie'd lay in bed cryin'. It was really hard on her. You know, here's her youngest fishing, and she knows he's a good kid. But everybody's sayin' how terrible fishermen are: "Is *that* what he does?" It hurts.

And the day the net ban passed, well, that was a really sad day. This all has dominated our lives for so long, we kind of had to get out of it. Just talkin' about it, it brings up all those bad memories. We pretty much learned to live with it. But let it go? You can't let it go. It's like sayin', "Let your heritage go, let your life go." Everything you've worked for since the time you can remember. It's givin' up your life, literally.

I sure hope that I'm fishin' when I die. And I hope my son's still commercial fishing. It don't look good. See, even offshore, the secretary for the Mississippi CCA is one of the sport representatives on the Gulf of Mexico Fishery Management Council.

The Gulf Council, which manages our fish in federal waters, it's slanted horribly. They're going to get what they want, whatever they want. What's goin' on here right now, if they take the redfish from us, which they're probably gonna do, it's going to put us on the verge of not fishin'. David may not be a full-time commercial fisherman if they take redfish. Unless we come up with somethin' else.

You've got to have somethin' in the winter. That's when we catch redfish. There are so few fishermen now, commercial fishermen, that quotas are not that big a problem. The quota used to get caught but now I don't think there's enough fishermen to fill it. But if they take that 40,000-pound statewide quota from us, it's gonna be bad, real bad.

They're workin' on it big time. That's CCA's drivin' thing right now in Mississippi: Gamefish status for redfish. But, shit, they want *everything*. I mean there are 150,000 sport fishermen allowed to catch three redfish a day. An average redfish, let's say four pounds, that's 12 pounds a day, per fisherman, times 150,000, times 365. That's *their* quota. And they want to take our 40,000 pounds from us. And deprive every consumer in Mississippi of this fish that is as much a guy's in Yazoo City as it is the CCA president's. He has just as much right to those fish.

But they're gonna get it because they get everything they want. They'll say, "Well, they've done it everywhere else, Texas, Lou'siana." They did that when we were in the net ban. "Is Mississippi gonna be last?" They played on the fact that we're a backwards state. "Are we gonna be last again? Alabama's doin' it, Lou'siana's doin' it." All these CCA pukes got up there and told them, "We're gonna be last, *they're* gonna do it."

Well, Alabama's fishin' full wide open. They never did it. But I don't know if that was all just rhetoric, you know? They were working behind the scenes. We know for a fact that some of the senators in Mississippi were getting cash payoffs, aside from huge campaign contributions. I heard of one senator getting a briefcase with a thousand dollars in it. Cash, no questions asked. But hell, what's the difference between that and a campaign contribution?

Who's sufferin' though? Who's gonna continue to suffer? The state economy, the Mississippi state economy is in dire straits right now. If we had the seafood industry that we had before the net ban hysteria, Mississippi would be in better shape.

I can remember when slabs [offshore shrimp boats] were tied up six abreast at the docks around here. Now it's rare to see out-of-state slabs come in here anymore, unless they come in to get their bottoms painted or something.

When Ingalls, the shipyard, would have big shutdowns, or temporary layoffs, everybody'd go to fishin', as a backup. All that's gone.

The sheephead fishery's gone, pretty much, a little bit left. The mackerel fishery is extinct, Spanish mackerel, gone. And that's what I made the majority of my livin' with. Drum, redfish, it's all just collapsed. It's gone. You can't catch them. You're not gonna catch fish in river water with a cast net when you can't see them.

Now, there's only three of us left here in Pascagoula. And when you just have three people out there fishing in the hardest conditions, with all these things they've put on us, you're just not gonna catch 'em. It's gone.

During the mullet run, it's a lot less now than it was. There's just a handful of people that come out and throw nets.

The last 15 years I sold all my mullet to O'Sullivan Seafood. But this year I'll probably sell them to Clark if there's a price. But the cast-net roe mullet can't compete with a gillnet mullet. Because the gillnet mullet is a select size. When we throw our cast net, we catch the little white-roe mullet [males]. And if we don't get any money for the white-roe mullet, then we're not making any money.

I've heard of guys getting 15 cents a pound for roe mullet last year. Tops, for us was probably 30 cents. A lot of times it wasn't but 20, 25 cents. There was so much white roe in the cast net. And we can't raise the mesh size on the cast nets because if we do, all the fish are gonna gill in it. You're gonna spend all day picking little fish out of the net and throwing them overboard, which we're not gonna do.

Then you've got Alabama and Lou'siana that can still use gillnets for mullet. And Florida's cast-netters, but they've got bigger fish. And I guess the Carolinas are still gillnettin'. So, all these guys are settin' the prices by what they're catchin'. And we're not doin' nothin'. We quit in the middle of the season last year. Everybody. No price.

We went fishin' redfish and drum, trotline. And we get a lot of freshwater catfish too. You know, when the river floods, they come out. David and I caught one the year before last, a 71-pound blue cat. Man, that was a big boy. And we catch a *bunch* of 30-pounders. And our average fish are about eight pounds. It's a lot of fun, it really is a lot of fun.

We know lots of places where the fish are as thick as flies, and you never see a sport fisherman. That's what's saving us—we stay away from them because we know where the fish are, and they don't.

And it's a night fishery. You run your lines out in the evening and the next morning you pick up all your lines. Then you bait them during the day and put 'em back out at night. So they don't see us doin' it. That's why we're still doin' it. We already use circle hooks, you know, but they would think of somethin' else. In fact, they already proposed to the Legislature of Mississippi that saltwater trotlines be limited to three hooks when, in fresh water, you can fish a thousand hooks if you want, for catfish. Three hooks! I mean that's just how absurd it all is. I wonder how many hooks are in the water from the sport fishermen?

When I worked for the University of Georgia, they had a documentary at Skidaway that they would show in the schools. It showed them gillnettin' in Lake Erie and the sports were trying to stop them. It's a wonderful tape. It shows them pullin' in the nets and talkin' about how they're having to band together, how the commercial guys have always been rivals and now they've got these sport fishermen to contend with who are tryin' to put 'em out of business. A perfect analogy to what we went through. And this was 1973, before I ever imagined that they would try to ban gillnets in the South.

And yeah, that's right, on the Canadian side o' Lake Erie they're still catchin' like 20 million pounds of fish a year. A different management system? Well, fishery "management" is a terrible word to use. It's not management, by any stretch of the imagination. It is allocation. And here, it's allocated to the recreational group.

Why do the politicians go along with it? "Because the CCA has put a bunch of money in my pocket," that's essentially it. "They took me out on their yacht. And they gave me a big campaign contribution."

Of course, you're not going to hear that from them but that's exactly what happened.

A lot of people really don't care, but I think it's more that they don't know. Because your average consumer, if he knew what was going on, that he was going to have to eat some fish that may have been urinated on by some Vietnamese before it was shipped over here, that's been frozen for a month, when he's got a fresh fish that can be caught today, and he can eat tonight. If he knew that that was replacing the fish that he has as much a right to as anyone else, they would be up in arms. They don't know all that. But it could change.

I've said it for twenty years, ever since the net ban thing started, we are too dependent on foreign food. And when these terrorists, as has been in the news lately—they think our food source is a threat. When they start tainting our food, then something's going to happen. I think that's the only thing that's going to save us. When they ship a load of fish or shrimp over here that's been raised in some cesspool in Vietnam or Ecuador or whatever and destroyed 20 percent of the world's rain forest doing it, and they put some poison in that fish and a thousand Americans die from eatin' that fish, then, maybe, they'll realize that we can walk on fish that are virtually untouched. I can't see anything else that could save us.

Because the U.S. government is so corrupt that it is just horrible. Immoral, corrupt, this country is really in bad shape. It's wonderful in a lot of ways but corrupt and immoral. Look at how it was when we were kids, compared to what's going on now. It's amazing what a turnaround.

I won't say "most," but a lot of the biologists are biased. That's where it started here. When they went to shut down the islands, there was a biologist at the NOAA facility here that was the spokesman for banning the islands. And he didn't have a clue what was going on. I called him up and told him off, wrote him letters. He was just one of these dreamers, hooked up with some of these people that didn't like us. Really, he was kind of stupid, he kind of fell for it. So he wound up with his name in the paper which made him my target, anyway. At that point, the other fishermen didn't really know what was going on.

That was the first thing that ever happened to us, the very first thing. Then they started the size limits, then they started the shutdowns. I wish I'd have listed it as it happened because it was just, this month this, this month that. Floats on the net had to be so far apart, and no scientific basis for any of it.

We begged them during the net ban, "Use the scientific data." Even if the data was flawed, even with their inadequacies, their data was on our side. And I'm sure it was the same way in Lou'siana. It's all on our side, but they won't use it.

A good story, in that respect, when Papa was a fisheries methods and equipment specialist for NOAA, the federal government did a big study on shrimp in the Gulf of Mexico. And spent hundreds of thousands of dollars on it. When it was all over, I don't know if it was in Washington or where it was, but after they did their study, everybody involved in the project got up and gave their take on it. And the biologists went "Oh, this was so wonderful, we got all this data, it was a great success, and this money was well spent."

Well, Papa got up there and he said one thing, "We haven't learned a damned thing that the shrimpers haven't known for years." He said he was not popular when he stood up and said that but that's the way it was.

And now they're spending billions of dollars, I'm sure the federal government is, on research when all they would have to do is get a handful of commercial fishermen and they could tell them all of it. They go looking for this stuff, but they will not use a man that has spent his life studying it.

That's like, now it's not to sound egotistical, but these turtle studies we do? They looked for Mississippi red-bellied turtles for 15, 20 years. And in that period, I think they caught seven. So I get a call from the

Mississippi Museum of Natural Science, and they told me what they wanted to do.

The first afternoon they got here, I said, "We'll go give it a try." We went out and caught five, and I showed them where they were nesting! We just blew 'em away. They love us—Scott, David, and me. It's not that we're so great, but we've been there—we've walked the marsh, we've walked the woods, we fished the waters. We know what's there, and where it is.

And if these fisheries organizations would use the commercial fishermen, think what they could know. But they'll go spend a million dollars on something that they could spend, probably, $20,000 on, with one commercial fisherman on the boat, and get the same data. Better data!

Yeah, we've had to do different things. When the Nature Conservancy would buy some land, they'd give me a contract to go in and catalog the reptiles and amphibians that were there. I've got all kinds of traps for that. I did several studies for the museum, but there's no money for that now.

So I got into building birdhouses. Not to brag on them but they're unique. I've been building birdhouses out of driftwood for three or four years now. And it's turned into a pretty good business. We sell them at these little arts and craft shows.

And this is all marine litter. I always figured I could do somethin' with it—those old worm-eaten boards that are thrown out. And it's kind of like fishin' except better—you're beachcombing! I'd rather beachcomb than anything.

I sold one birdhouse for $150 at the last show. I didn't want to sell it, really, I just put it in my booth for show. I figured that nobody'd pay $150 for that thing but shoot! That was at a show in Ocean Springs, and we sold every birdhouse I had but two. But I didn't realize what people

Pete Floyd created a niche business for himself building birdhouses
out of driftwood.

would pay for them. I was charging like $25. And then the next show I
jacked the price up. And it made no difference whatsoever.

Now that I'm learning, I might charge even more. But you've still got
to keep the less fortunate people, you've got to give them access to it.
You get these people that obviously don't have any money and they're
just going on and on about how neat that is, and you can just see in their
eyes that they want it so bad. I don't have it in the heart to just let 'em
walk away. I drop the price for people like that. If I'm wanting $40, I'll
say, "Well, okay, you can have it for $20."

The bottles? I found a real good one a couple weeks ago—an 1865
Blatt's Beer bottle, mint condition. Boy, it's a nice one.

I know where there's a site where a homestead was, from about 1830
up until a hurricane took it out. It was where the, well, I don't want to

179

talk too much, but there was an old lighthouse out there. The hurricane took it out and I think it killed the whole family—the lighthouse keeper and something like seven girls. They tried to get him to leave but he wouldn't.

Well, I found the damned place. You would not believe the stuff we've found there. I always go there after a storm because more washes out. As a matter of fact, when we see a hurricane's comin' here, we get our gas to go to that site before the storm hits. And that's the first thing we do before we clean our yard or anything. We go to that site.

I found a couple of amphoras there, I found two links of a gold money chain. My son found two 1860 coins. You just never know what you're going to find. And bottles, I mean it's just the Holy Grail for bottle collectors. The stuff you find that's been buried in that sand, it's just in mint condition—old and good.

Did I ever sell any? I sold one. There was a piece of land I wanted up on Bayou Cumbest and it came up for sale for $10,000. My wife threw a fit, she didn't want me to buy it. And I said, "Man, that could be St. Augustine, Palm Valley. That may be worth a damned fortune one day."

She was still against it, so I said, "Well, I've got a bottle."

And I sold one for $3,500, came up with the rest and went and bought the piece o' land. I hated to sell the bottle, but you can't drop that piece o' land and break it.

We're surviving, so far. We're not gettin' rich. But have you ever known a fisherman to get rich? We're trying to stay in the boat, in the marsh or the water, to make a livin'. We're still doin' it.

I've been real fortunate—the things I love to do the most, I get paid for. You know, my hobbies—turnin' beachcombin' into birdhouses, turnin' catching snakes for fun into research surveys, and of course, fishin'.

A long shot. David Floyd throwing cast net for spadefish in the Gulf of Mexico.

But fishing's not a hobby, I don't know what you'd call fishin' in our family. A way of life? "Way of life" is not strong enough. It's just something that *is* the Floyds. Fishing is Floyd, Floyd is fishin'.

A good way to put it, I know you've heard, "Saltwater runs through your veins," well, we're part of it. We're an animal that lives in the ocean, is a good way to put it. We're just like a porpoise or anything else.

And there are thousands and thousands of "Floyds" you know? They might be named Stork, or they might be named Bosarge or, I don't know what the commercial names are in Lou'siana. But if you take us out, it's like taking out an osprey. We're part of it.

See, I told all of them, my family, during the net ban, "They're not gonna stop a fisherman from fishin'." We're gonna do something else. We're gonna get cast nets, and if they ban the cast nets, if they ban the trotlines, we're gonna catch them somehow. We're gonna catch fish. We're fishermen. I don't care what they do. They're gonna knock us financially but they're not going to stop us. They can't stop us unless they kill us.

I'm still at the fire station. I work one 24-hour day on, and two days off. When I get off, David and I usually hop right in the boat and go. When I'm at work, he goes off on his own, running his crab pots or whatever.

Back when he quit college to go fishing, he told his mom that there was just no job that he could get with his education that he would like. It hurt Jeannie because the two older kids had their degrees, but I told her, "Well, hell, that's what I did."

You know, when I was two and a half, I was on the boat with my daddy, literally. And my sons were on my boat with me when they were two and a half, spending the night. Just me and the little boy. Nowadays, if you did that, people would think you were crazy.

But I can remember Scott when he was little, we were out in a Bosarge skiff—that's just a little open 18-foot boat, about 5-foot wide, wooden, with rails about 18 inches tall. Scott was so little that he would have his chest against the rails of the boat, pullin' in sharks on a handline. You know, I'd have the gillnet out and I'd have some string and a hook, and I'd throw it out for him.

Right now, we're cast-netting for spadefish, around the offshore oil rigs. As far as I know, there are two other people doing it out of Pascagoula. We just started and we're still learning. We fish in 40 or 50 feet of water, and you've got to let the net sink all the way to the bottom before you bring it up. That's rough.

We're getting 40 or 50 cents a pound. Spadefish have got a meat that's kind of grayish when you fillet it, but it cooks up real white. They ship them to New York and California. In California, they call them angelfish,

(Opposite) Pete Floyd icing down a catch of spadefish.

and it's in the same category as an angelfish, but it's probably on the lower end. It's been around for a long time—they seine 'em and hook-and-line 'em on the East Coast. But we never really messed with them here before.

We threw on a bunch of fish one day that we didn't even know what they were. And neither did the fish house. But they bought 'em! That's how it is here now. People are hungry for any kind of fish.

Another offshore fish that we're trying to learn to catch is blue runners. We've had days when we'd get like a hundred pounds, maybe 150 with the cast nets but that's the exception. If you could find out how to catch those damned blue runners, you could get rich. The buyers here and in Alabama, all the big shippers want the blue runners. At 50 cents a pound, that's not bad money, and it's unlimited. I think they go to California but I'm not sure. Either California or New York, but they want all you can get, anytime you can get them.

Offshore, there are just jillions of 'em. You see 'em feeding up on top all the time. We're gonna figure out a way to catch 'em. And when we do we're gonna be happy.

LAST MAN STANDING

It is an issue that has been ongoing for over two decades now. When I researched how the current regulations came into being, I realized that this was a politically charged issue and the regulations of 1996 were enacted based on the political tide at the time, instead of on the tenets that we have sworn to abide by when making fishery management decisions. This is not 1996, and we are not that commission.
—Steve Bosarge, Mississippi Commission
on Marine Resources, 2014

Rod and reel don't fit in my hand. I was raised to fish with a net. Why can't I do that?
—Richard Gable, commercial fisherman, 2014

The fairest way to do commercial fishing is have everyone use the same gear. ... Have the guy with the haul seine switch to a rod and reel and participate with everybody in catching those fish. Please do not be unfair.
— Steve Shepard, Sierra Club, 2017

The one thing we cannot control in the recreational fishery right now is effort. We cannot do it, we do not have the ability to put a moratorium on licenses, or anything. We are going to continue to see more recreational fishermen, and they are mainly going to target the sport fish such as spotted seatrout, red drum and southern flounder to some degree.

—Matt Hill, Finfish Bureau Chief, Mississippi
Department of Marine Resources, 2017

We strongly encourage the Commission to pursue true conservation measures such as implementing an annual total allowable catch in the recreational sector.

—Ryan Bradley,
Mississippi Commercial Fisheries United, 2018

I'll tell you, it's hard to squeeze a few dollars out of what's left, no doubt.

—Martin Young, haul seiner, 2024

CHAPTER ONE

2010

REALITY CHECK:
RECREATIONAL LANDINGS UP,
COMMERCIAL LANDINGS DOWN

From the mid-1990s—when the sportsmen first went on the attack—until 2010, their annual haul naturally ballooned as more people went angling and they inhaled some of the fish that commercial fishermen could no longer net.

In our benchmark year 1994—the last before the net-ban turmoil began—recreational anglers made an estimated 2.7 million trips in Mississippi's salt waters. By 2010, that number had increased 65 percent, to more than 4.5 million, and landings of the anglers' 14 most popular inshore species increased 50 percent, from nearly 4.5 million pounds in 1994 to nearly 6.7 million pounds.

In descending order of volume that 2010 catch included spotted seatrout (2,250,514 pounds); sand seatrout (1,014,725 pounds); red drum (977,967 pounds); southern and Gulf flounders (690,516 pounds); black drum (584,445 pounds); striped mullet (332,715 pounds); sheepshead (326,629 pounds); Atlantic croaker (240,806 pounds); southern and Gulf kingfish (168,323 pounds); gafftopsail catfish (39,086 pounds); Spanish mackerel (19,608 pounds); and pinfish (12,452 pounds).

Meanwhile, commercial landings of the same species tanked.

In 1994, the seafood producers had put nearly 1.5 million pounds of inshore fish on the market, with a dockside value better than $860,000.

By 2010, their harvest of the same species had dropped about 85 percent, to 211,570 pounds worth $282,264.

In descending order of volume, the 2010 landings and dockside values included striped mullet (58,796 pounds, $30,552); spotted seatrout (41,534 pounds, $96,044); red drum (36,444 pounds, $64,619); flounder (27,565 pounds, $64,101); sheepshead (24,852, pounds, $14,665); black drum (11,442 pounds, $4,151); sand seatrout (8,042 pounds, $5,225); and king whiting/kingfishes (2,895 pounds, $2,907).

(Landings of Florida pompano, Spanish mackerel, Atlantic croaker, bluefish, ladyfish, spot, tripletail, and sea catfishes weren't available because they were purchased by a single buyer and therefore considered to be "confidential." Without nets, the commercial harvest of these species would have been minimal.)

While the volume of their harvest plummeted, fishermen generally received more money for the fish they brought in: Spotted seatrout in 2010 brought $2.31 per pound, up 62 cents from the 1994 average of $1.69, and red drum were up 55 cents, from $1.22 in 1994 to $1.77 in 2010.

Mullet prices stagnated at 52 cents: While Mississippi's production declined—from 781,300 pounds in 1994 to just 58,796 in 2010—catches of roe mullet in some other states and foreign nations soared. So local dealers lost their pricing power and had to take whatever they were offered.

• CHAPTER TWO •

NEARLY NYLON

Gulfport fisherman Richard Gable had tried as hard as anyone to make it with gear that met the degradability standards the Commission on Marine Resources had imposed in the 1990s. In 2003 and 2004, he was catching some fish with trammel nets that were made of linen, mostly.

When a redfish or drum hit his net's small-meshed webbing, it pushed it through a "wall" of 14-inch-mesh squares that hung on either side, and neatly bagged itself in a pouch. Trammel net walling was unavailable in linen or cotton yet most of the suppliers stocked it in nylon. So that's what Gable used.

After all, he reasoned, it was the small-mesh webbing that caught the fish, not the panels that hung on each side with meshes too large to stop a basketball. He also figured that there were plenty of other components in his nets that weren't biodegradable including the hanging twine (nylon), ropes (polypropylene), floats (polystyrene), and lead weights. The Department of Marine Resources had even okayed his gear with a tag.

Still, Gable was spending a thousand dollars to build a net that lasted only three months. Then his source of natural-fibered webbing dried up. By 2012, he'd had enough.

When he told the commission, at its December meeting, that he couldn't find any degradable webbing anywhere, the panel directed marine fisheries staff to assist him. The agency did locate some Chinese-made cotton webbing in Tampa and Gable bought 90 pounds, he explained to the commission a year later, in January 2014.

But the net was more than his crew could handle. "Me and my wife fish. And she ain't no bigger than a minute," he said. "And for us trying to handle a 23-foot-deep net in cotton that's rotting away just does not make good logical sense."

The material's tendency to decompose was Gable's main objection: "Every time you use that net you got to take it and hang it. I put it inside of our shop which is 30- by 80-foot long," but the cotton absorbed moisture from the air, said Gable. "You can go out there in the morning and feel it, and it's wet. And, by it being wet, it's steadily rotting.

"Now, the money I had to spend on this, it was $10 a pound, this is a 32-pound bundle—$320 basically went to poop. That's what I'm being made to fish with," he said, as the commissioners tugged at samples of the deteriorating material.

It didn't "make good sense" for him to have to buy 90 pounds of webbing from China "whenever there's a man in Biloxi that has a nylon company in Ocean Springs that I could buy my webbing local," said Gable.

"Rod and reel don't fit in my hand," he added. "I was raised to fish with a net. Why can't I do that?"

Gable knew that efficient monofilament webbing would be out of the question but didn't think it unreasonable that he be allowed to make his gillnets with untreated multifilament nylon webbing which was not only rot-resistant but readily available in a variety of mesh sizes. Mindful that the process associated with such a regulatory change would take months, he also made a simpler request that he thought could be quickly granted: Allow him to once again add nylon walling to his cotton nets. (After Hurricane Katrina, in 2005, the DMR inspected all the finfish nets then in use, and disallowed Gable's hybrid trammel nets.)

With government, however, nothing is simple, although commissioners were surprisingly sympathetic: "We welcome ... honest and sincere comments. And yours have been that way," Jimmy Taylor, the panel's charter-boat representative, told Gable. "And that's the reason why the CMR has tried to accommodate you. We know that you're speaking from the heart, and you're a hard-working person."

Gable's written requests in fact served as "petitions for reconsideration," which forced the commission to either grant or not grant him the relief he requested. Over the next several months, the commissioners, along with the agency's staff, and with plenty of input from the public, crafted a regulation that would—if approved—allow Gable and a handful of other eligible net fishermen to finish out their careers with nylon webbing.

DIEHARDS

Over the years, the Department of Marine Resources had certified a total of eight gillnets and trammel nets as degradable. Not all those nets were in use: "As best we can tell, anywhere from three to four of them are active," said finfish bureau chief Matt Hill.

Trip-ticket data indicated that, in 2013, those diehards had netted less than two thousand pounds of fish, which "ranged from striped mullet to southern kingfish, a handful of red drum and speckled trout," said Hill. "In 2012, it was less than that. I don't believe it even added up to be a thousand pounds, in the gillnet fishery."

Hoping to stem even further decline, the commission's commercial fishing representative, Steve Bosarge, led the charge for reinstituting the use of nylon webbing.

A third-generation fisherman from Pascagoula, Bosarge was CEO of Bosarge Boats, a family business that ran a fleet of seafood harvesting, fisheries research, and other specialized types of vessels in the Gulf and South Atlantic.

When he asked DMR biologist Matt Hill if reintroducing nylon webbing would create "a drastic, crazy problem," Hill replied that the limper, multifilament nylon was "an extremely difficult material to use," compared with monofilament. "A lot of guys will probably be one and done with it, to be honest with you."

For good measure, any new regulation should ensure that only the "right people" be allowed into the fishery, said Hill, and "have ways to get people out of the fishery that are repeat offenders, so it does not ruin it for the rest of them."

PUSH BACK

Recreational fishermen weren't convinced. They were afraid that, like the camel's nose under the tent, a little nylon today could lead to a lot more tomorrow.

"I'm hearing the motion that you are going to actually have these really, really honest gillnetters and only really, really honest gillnetters gillnetting," said Ocean Springs sport fisherman Steve Shepard, a CCA officer and chair of the Sierra Club's Gulf Coast Group. "The only really honest gillnetter I ever knew passed away," he said. "The rest of them would put the net right around the wharf right in front of you."

The Sierra Club objected to weakening the biodegradable rule because it would "open up a lot of activity, a lot of fishing activity," Shepard told the panel.

Furthermore, gillnets were clearly not needed, he said, because the seafood markets remained stocked despite the virtual elimination of the gear: "We do it with cast nets and hook and line."

Like *Florida Sportsman*'s Karl Wickstrom, Shepard wanted sport and commercial fishermen to use identical tools "The Sierra Club believes that the water should be equal to all—not ultimately perfectly equal—but I do think everybody ought to have the same equipment out there,

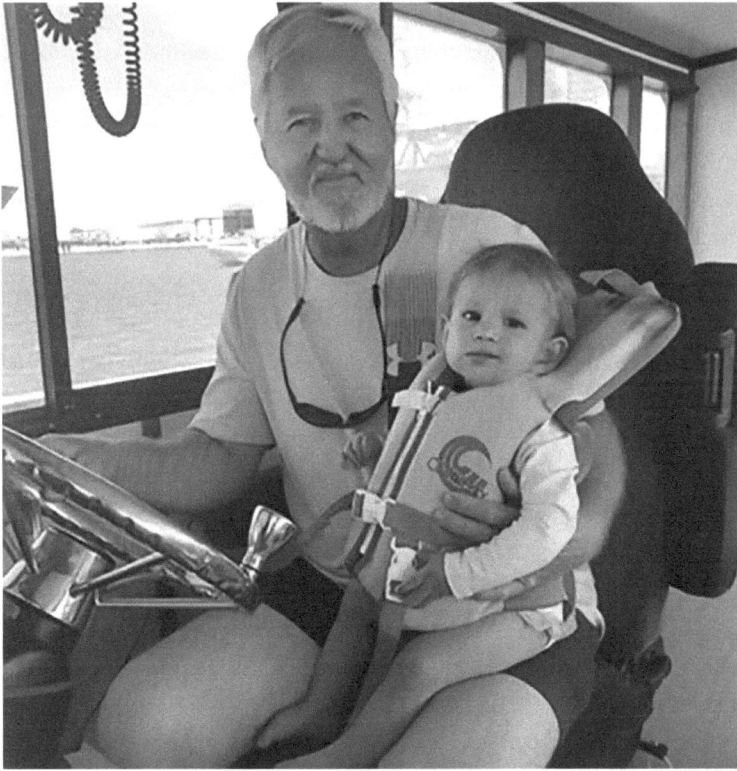

Steve Bosarge and deckhand. *(Courtesy of Bosarge family)*

and then, the skilled ones can sell the fish because they catch more than they can eat, and the unskilled ones have a chance to catch fish."

Johnny Marquez, the Mississippi CCA's executive director, reminded the panel that when the issue was "really in the heat of things back in the nineties, the degradable net material was entered as sort of a last-minute compromise, rather than banning the nets entirely, and it has been a very effective provision, as far as we are concerned."

Still, his group would prefer to see "some regulations where we, perhaps, ultimately sunset this gear, perhaps provide some relief to the guys that are fishing right now but see a retirement of the ... entanglement-net gear over time."

What the group did not want to see, said Marquez, was "a change to some non-biodegradable material that then encourages others to come back into the fishery."

EXPERT ADVICE

The 1996 ruling of the Commission on Marine Resources had permitted the webbing of saltwater gillnets and trammel nets to be constructed of untreated cotton or linen only. Any new candidate would have to lose at least 50 percent of its tensile strength after one year's immersion in salt water and be identifiable with a field test.

Or the commission could simply add a substance to its short list of approved materials.

"What we aim to do here is to, basically, change the restrictions and allow nylon," said commissioner Steve Bosarge. "That is what I would like to see."

For a broadened perspective, the agency turned to the academy where, in a sense, the biodegradable conundrum had originated.

Jeff Wiggins, Ph.D., interim director at the University of Southern Mississippi's School of Polymers and High-Performance Materials, had been studying the hydrolytic degradation of polymers for over twenty years, so DMR staff wrote him and asked, "Do you know of any materials currently available that would meet the standards in our regulations?"

"This is a very difficult question," he responded.

The "technical" answer was "yes" because medical sutures would meet the standards, he said. But the "practical" answer was "no" because suture material degraded too quickly in water and would be prohibitively expensive.

"Economics have to be a factor in the fishing community," he explained.

"I'm guessing this is an unsatisfactory response," the professor concluded. "There could certainly be some new technology I am not aware of, and I know this issue for fishing applications is not new. The scientific community has been aware of this quagmire for decades, and I have not seen a reasonable solution to date."

In short, according to the professor, there weren't any materials that met the commission's standards. In that case, why not approve something else?

PUSHING THE ENVELOPE

The amendment that the commission was considering would allow fishermen who'd possessed a net that had been inspected and tagged by the Department of Marine Resources between September 1, 2005, and March 18, 2014, to use webbing of uncoated multifilament nylon twine.

"Now, if all were … trying as hard as myself, then maybe they should be part of it," Richard Gable told the commission at its May 2014 meeting, when the matter was to be finally resolved. "But the ones who have the tagged nets should be the only ones allowed to fish with the nylon—no one else. When we die, it's over. That's not right, but they don't make real commercial net fishermen anymore," he said.

The handful of qualifying fishermen would be permitted to use nylon webbing until they retired. New entrants wouldn't be barred from entering the net fishery, but their webbing could only be made of uncoated cotton or linen.

To ensure compliance, the DMR would annually inspect each commercial net and affix a tag with a unique identification number.

In an attempt to weed out some habitual offenders, the licenses of anyone who received three fishing citations within any three-year period would be revoked.

Gable agreed with the three-strikes-and-you're-out provision: "If the fisherman is continuing to do something against the laws or the gillnet rules, then, if he does it three times, he ain't got no business being there. And I think if it was done that way back in the nineties, we wouldn't have had the trouble that we have now."

As for the issue of "ghost nets," a "true commercial fisherman will not lose his net. You just don't do it. There's too much money in it," Gable told the panel.

"Now, some has left nets and never picked them up, just like a drunk driver [that breaks the law], but not all do it. … A real fisherman is not going to lose his net. He's not going to leave it unattended. He's going to abide by the rules and regulations. I think out of 30 years of fishing, I got one ticket because I was in an unmarked channel … because it was too rough to run out front. Actions speak louder than words," he said.

The commission's seafood processing representative, Richard Gollott, interjected. "I think it is ridiculous that we are even talking about this this much. We're only talking about eight people. The majority of these people are over 70 years old. It's basically a dying fishery. It's ridiculous to put your foot on somebody that's dying already."

"It's not a dying breed. We make it that way," objected Pascagoula net fisherman Larry Ryan Jr. "This is not right, and it's never been right."

Ryan had grown up in the nineties, he said, "when it was a battle and it was a fight to watch my parents, my uncles, my aunts, my family get degraded time and time again over somebody's opinion."

Nylon would make it easier for him to cull his catch and to obtain webbing of the right size "at the time I need it," he said.

As the vote neared, the panel's commercial fishing representative made an impassioned plea to his fellow commissioners. Bosarge didn't just want the handful of qualifying fishermen to be able to use nylon until they retired, he wanted their permits to be transferrable to new entrants who could continue to use the material.

Giving eight people the opportunity to use nylon gear that went out of date back in the 1970s wasn't "giving them the bank," he said. "In other words, we're just trying to right a little bit of a wrong."

The netting issue had been ongoing for more than two decades, Bosarge explained. "When I researched how the current regulations came into being, I realized that this was a politically charged issue and the regulations of 1996 were enacted based on the political tide at the time, instead of on the tenets that we have sworn to abide by when making fishery management decisions. This is not 1996, and we are not that commission."

In a 1997 law that was promoted by Jean Williams and her seafood industry group, Mississippi's legislature required that the Commission on Marine Resources follow the same guidelines in managing fisheries that the U.S. Congress had codified in the federal Magnuson Act. They included phrasing such as:

- Conservation and management measures shall be based upon the best scientific information available.

- Allocation shall be fair and equitable and carried out in a manner that no particular entity acquires an excessive share of the privileges.

- Measures shall minimize adverse economic impacts on fishing communities.

"Now we must apply the above evaluation criteria to the regulation currently on the books … in order to determine if this amendment is appropriate," said Bosarge: "Does the current regulation ensure that any allocation of fishery resources is 'fair and equitable' and 'reasonably calculated to promote conservation?'

"No," because it had resulted in a "de facto reallocation from the commercial sector to the recreational sector," he said. "All fishermen that harvest our marine resources should be held to the same standard. We cannot allow one group of fishermen to use monofilament and Spectra, knowing the detrimental effects it has on the environment and marine mammals, and then not even allow the other group of fishermen to use nylon.

"We cannot allow an open and unlimited-entry fishery to one group and then tell the other group, 'I'm sorry but we are going to put a moratorium on you and then we will make your license non-transferrable so that you will be completely eliminated from the fishery.'"

If the panelists didn't enact the amendment to allow nylon, they would be "in clear violation of our legal mandate as fisheries managers," he said. "Not only is it clearly the right choice based on the scientific data, but it is morally the right thing to do," said Bosarge, who ended his appeal, "Anyone who walked into this room today ... with an open mind that listened to the facts presented to the commission could come to only one conclusion—that the current net regulation should be amended to allow these eight remaining fishermen to use nylon webbing and to have a transferrable license."

VOTE

Bosarge moved to accept the new regulation as it was originally written, which limited the use of nylon to retiring fishermen. After that passed, he hoped to "come back and do the other thing," which was to amend that rule to allow the retiring netters to transfer their nylon permits to new entrants.

The measure needed a majority of the panel's five members to vote in its favor.

(When the commission was first authorized in 1994, it had seven members. In 2002, the Legislature reconstituted the panel by eliminating the slots for a non-seafood "at-large industry" person and the coastal Fifth Congressional District's sitting representative on the Commission on Wildlife, Fisheries and Parks.)

Bosarge and Biloxi shrimp processor Richard Gollott voted in favor of the amendment; Hancock County's Ernie Zimmerman, who succeeded Vernon Asper as the panel's environmental representative—through his memberships in the CCA and Mississippi Wildlife Federation—voted no. So did Jackson County's Shelby Drummond, the panel's private recreational fisherman.

"And so we have a tie," said Chairman Jimmy Taylor, who ran a billfish charter out of Biloxi. "And the chairman votes no. The motion fails for the passage of Title 22 Part 5.

"What it means is that the law will remain the same as it was before we brought this up."

WHAT IF?

Had Mississippi's Commission on Marine Resources treated each of the state's fishing sectors "fairly and equitably" back in the mid-1990s, it's possible that all the commercial nets and recreational fishing lines on the coast would now be degradable.

In a May 22, 2012, note to New Moon Press, publisher of this book, a representative of Jarden Applied Materials, a division of Shakespeare Monofilaments, confirmed that his company had offered a degradable nylon monofilament back in the late nineties. The formulations of Shakespeare's DN-series could be adjusted to "tune in the desired degradation life of the material," he said.

The company had partnered with a fishnet company to produce nets that were put into commercial service on the West Coast, he said, but the company discontinued its degradable fishnet program due to a lack of demand—there weren't enough laws mandating its use, and in the "extremely price-competitive" monofilament fishnet market, it was much more costly than conventional nylon monofilament.

Commercial netting is fabricated of monofilament lines of varying strengths according to the size of fish being targeted. Had recreational anglers been forced to adopt degradable lines, the increased demand for the material could have made its production profitable.

PLASTIC IN THE ENVIRONMENT

The Ocean Conservancy, the nation's largest marine-focused nonprofit, had sponsored an annual coastal cleanup since the 1980s (when the group was called the Center for Marine Conservation). A report on its 1995 International Coastal Cleanup described how lost or discarded monofilament fishing line could injure or kill marine mammals, sea turtles, birds, fish, and shellfish:

"Once entangled in the line, ensnared animals are usually unable to free themselves and can eventually become exhausted and drown; the strong yet thin line can cause abrasions that become infected; or the entangling line could impair the animal's ability to catch food, avoid predators, or may become caught on branches, power lines, or submerged structures."

Between 1989 and 1995, monofilament debris was the leading cause of animal entanglements: Of the 647 incidents recorded in that period, 227 (35 percent) were attributable to monofilament line.

In the 1995 cleanup alone, international volunteers collected 53,861 pieces of plastic fishing line and plastic floats and lures.

Photos on cover of the 2011 Mississippi Monofilament Recycling Program brochure show a seahorse and osprey entangled in line, and a dolphin trailing line from a barbed hook in its mouth. Photos inside show a sea turtle and mangrove snapper also trailing monofilament line from embedded hooks.

In 2008, Mississippi's Department of Marine Resources initiated a monofilament recycling program to reduce the amount of line anglers intentionally discarded into the environment.

At the commission's November 18, 2014, meeting, the DMR's Wes Devers told the panel that monofilament was especially hazardous for birds. "They tend to bring it to their nests, and they get tangled in it, and then, obviously, they will starve to death because they can't get away from it," said Devers, who explained how mono could be hazardous for humans as well: "If somebody is diving on a reef and there is a lot of line down there, it can get tangled up in the regulators," he said.

"It is just good to keep this out of the water."

From the agency's 2011 brochure on its recycling program:

- Monofilament is a strand of strong, flexible plastic used for fishing, and the majority of it is non-degradable in water and lasts about 600 years!

- In the 2010 Mississippi Coastal Cleanup, 261 pieces of fishing line were picked up and recorded in just three hours, and two fish were found entangled in monofilament.

- Fishing line was the No. 1 entanglement item during the 2010 International Coastal Cleanup (ICC), and over the ICC's 25 years, fishing line had been the No. 1 entanglement item of all time, with more than 40 percent of all animal entanglements.

In the first two years of Mississippi's Monofilament Recycling Program more than 254 pounds of fishing line was collected in about 35 stations.

CHAPTER THREE

THE GILLNETS ARE BACK!

In 2014, when netter Richard Gable was trying to convince the commissioners to let him use nylon webbing, he told them that he'd not only had a degradable net "from the start" but he believed that he was the first one in the mid-nineties to have a seine. "They're allowed with no restriction except no longer than 1,200 feet," he said. "They can be nylon, mono, or whatever else."

No restrictions? Mono?

Real haul seines were massive, laborious to deploy, and pricey. At the time, the Biloxi Fish Net Company was charging about $6,000 for a ready-made one.

Unlike gillnets, which may be fished passively, seines have to be actively worked: After the net is run around a school of fish, or an area likely to hold some fish, the end is slowly dragged back to the starting point which herded the fish inside into a bag that looked like the tail of a trawl (See illustration in Appendix).

That bag—which enables the user to handle his catch in bulk rather than individually—is a distinguishing feature of a seine. So, when some fishermen in 2015 began to stitch "bags" to their old monofilament nets and show their new "seines" to the folks at the DMR, they had no problem getting them approved.

"Essentially, what they did, they found a loophole in the law," explained Ryan Bradley, executive director of Mississippi Commercial Fisheries United. "Since the haul seine wasn't defined, they were able to sew a nylon bag—that most probably wasn't even functional—onto a gillnet. It was 3- or 4-, 5-inch or whatever size mesh you wanted. Bring it down there and they'd say, 'Oh, this is a haul seine,' and hand them a haul-seine verification form."

In reality, these were strike nets or runaround nets, which are set around fish and hauled back right away.

"So much time had gone by that the law enforcement guys, they didn't know what the hell these guys were using. They were telling them, 'As long as you make a circle with it and you're not setnettin' with it, you know, leavin' it unattended.' That was the thing—they were telling them it had to stay attached to the boat and all this stuff. But in all reality—I'm not going to sugarcoat it—these nets were operating like gillnets," said Bradley.

On Mississippi's busy coast, it wasn't long before word got out: "The gillnets are back!"

"It took about three years for a fever pitch of CCA opposition to build up," said Bradley. "Eventually, it became really bad—the trigger point was in 2018 when we had a Bonnet Carré Spillway opening and it pushed all the dang trout up on Cat Island and these boys went in there with them nets and mopped up."

SPILLWAY

Louisiana's Bonnet Carré Spillway was constructed in the early 1930s in response to the devastating Flood of 1927. Located upriver from New Orleans, the spillway was designed to protect the city by diverting flood waters from the Mississippi River into Lake Pontchartrain.

Overflow from the lake funnels eastward into Mississippi Sound where salinities plummet. That fresh water, along with the outflow of the Pearl River, can wipe out the oysters on the Sound's western reefs. However, saltwater finfish like spotted seatrout are able to migrate to more

Ryan Bradley, a fifth-generation fisherman and owner of the Sea Alis Seafood Company, in Long Beach, was executive director of Mississippi Commercial Fisheries United, which took up the fight after Save America's Seafood Industry (SASI) faded away. *(Photo courtesy of MSCFU)*

hospitable locations, and after the U.S. Army Corps of Engineers opened the spillway in March 2018, they piled up around Cat Island.

Mississippi's commercial fishermen operated under a 50,000-pound speckled trout quota that was broken down into two seasons. The first season opened on February 1 and ran through May 31, or until 25,000 pounds were caught. The second season opened in the fall and remained open until fishermen landed their second 25,000 pounds.

"Well, these boys caught so much fish, they caught the whole 50,000 pounds in the first season," said Bradley. "It blew up, videos started leakin' out of these guys out and about and unloading at the dock. It just got bad to the point that recreational fishermen were all callin' their legislators. I mean they're burning the phones up. And we're hearing the flak."

"It had become a real heated battle, and the issue came up at our state marine resource commission. But instead of trying to say, 'Oh no, let's fix this gear that they really probably shouldn't ought to be usin',' they said, 'Let's just ban all the nets around Cat Island.'"

CAT ISLAND

Cat Island is the westernmost of Mississippi's four major barrier islands. It was discovered by the French in the late 1600s and after more than three centuries was still mostly natural.

The hammer-shaped island's interior supports a maritime forest of slash pines and live oaks ringed with marshes and bayous, dunes, and surf-washed beaches. Rich meadows of submerged vegetation grow in the island's lee waters, and the whole shoreline of this eight-mile-long fish magnet is etched with inlets, coves, and points that were ideal for hauling a seine.

Martin Young would know—he'd been seining around Cat Island for decades with a genuine, old-school haul seine.

"The haul seine survived the net-ban thing in the '90s thanks to me and my brother," said Young. "After they banned gillnets, most of 'em went to crabbin' and went and got jobs or whatever. But me and that brother of mine, we kept fishin' because that's all we'd done our whole lives. We was the only two that used seines for all them years since the ban."

The commission's 1996 ruling stipulated that gillnets and trammel nets had to be degradable, but it said nothing about haul seines. So, the Young brothers were able to work theirs, legally, for 20-some years.

Now, thanks to the sham-seine controversy, their gear, their best fishing territory, their livelihoods, were all in jeopardy.

THE GILLNETS ARE BACK!

Looks fishy: Cat Island lies seven miles offshore of Long Beach; this is its eastern end. *(U.S. Geological Survey photo by James Flocks)*

SEINE SCHOOL.1

When the Commission on Marine Resources took up the haul seine issue at its September 19, 2017, meeting, DMR fisheries director Joe Jewell provided some context.

With a map of the coast, he showed that haul seines and other finfish nets were already permanently banned from nearly all the state's productive waters. Additional areas, which were open to netting during the winter months, were closed during the summer. Those included Telegraph Key, one mile around Round Island, and one mile around Cat Island.

In another map, areas closed to recreational fishing were also shaded in green, said Jewell, and "you can see, there are no areas shaded in green which is to say there are no recreational closure areas for fishing."

The biologist also contrasted the harvests of the sport and commercial fisheries:

In the first six months of 2017, the two sectors boated a total of over 1.8 million pounds of the three "quota" species—redfish, trout, and flounder. Recreational anglers took 95 percent and the commercial sector took 5 percent. The "haul seine industry" accounted for 1 percent of the total landings, and hook-and-line and other commercial gear the other 4 percent. (At this time, the "haul seine industry" presumably included the few genuine haul seines that were in use plus the controversial "sham seines.")

Sport and commercial landings of all the most popular inshore species totaled about 3.5 million pounds in the first half of the year. Recreational landings accounted for just over 3.3 million pounds—93 percent—and commercial landings 7 percent. The "haul seine industry," again, accounted for 1 percent of that catch, and all other commercial gear 6 percent.

Could the "haul seines" potentially become ghost nets "like the old gillnets used to be?" asked Mark Havard, the panel's recreational fishing representative.

"These nets are very expensive to make," responded DMR biologist Matt Hill. "These fishermen are making a living. It's not the most lucrative living. I find it very difficult to believe that they would leave a $2,500 net in the water to ghost-net fish. The fishermen that are using them are very seasoned fishermen."

When questioned about the incidental catch of species other than those being targeted, Hill noted that the agency's enforcement officers had observed "haul seiners" making "very clean sets."

They were "very effective in targeting what we like to think of as underutilized species," which fishermen using hook-and-line or other

THE GILLNETS ARE BACK!

Martin Young, center, and his crew haul seining on Cat Island. Note winch, aft on the starboard side, used to haul back net. *(Ryan Bradley photo)*

gear types weren't interested in catching, said Hill. Species such as mullet, whiting, croaker, sheepshead, Spanish mackerel, ladyfish, didn't bring as much money as redfish and trout, but they were "still a valuable fish in several markets, and that is how these fishermen are making their living. It's not the quota species. It's the underutilized species that we are seeing."

The "haul seiners" could "target the exact species of fish they needed to catch, based on mesh size," Hill explained. "If they don't want to catch spotted seatrout, they will use a 4 ½- to 5 ½-inch mesh to where spotted seatrout will go through there, but they may possibly be targeting black drum, or they may use a different mesh size in a different area for Spanish mackerel," said Hill, who was in fact describing the runaround or strike-net fishery even if it was then being conducted as a "haul seine" fishery.

The confusion stemmed from the lack of a legal definition for the gear, which the CCA's F.J. Eicke hoped to remedy. Eicke, an Ocean Springs psychologist and chairman of the state chapter's government relations committee, said "This is about the definition of haul seines as they were observed by some of our members," who had contacted him with the message, "The gillnets are back!"

This is really a gill-net problem, he said, "that goes back to the 1990s when that particular issue was not exactly approached in a calm manner. We don't want to get back there. We want to do something that is legitimate and that allows commercial fishermen to function in a way that basically fits the history and current regulations."

Steve Shepard, of the CCA and Sierra Club, told commissioners that using nets with mesh sizes that conformed to the size of the fish was gillnetting. "That is entanglement, and that should not be allowed."

Shepard disagreed that fishermen would be unlikely to lose their expensive nets. "I know gillnetters. I know them, and they admitted it because, if they got into a mess of catfish or something, they dropped the net. It was easier for them to go home and get another net than it was to fight a big balled up mess of bycatch which, now, you say never happens. … If they ditch them, we are going to have ghost fishing again like we had back when the gillnetters were not responsible, and I don't see any reason why the haul seiners will be responsible. … They've got a track record in this state. It's not good."

It wasn't right that a few people got to use haul seines while everyone else was limited to fishing rods, said Shepard. "The fairest way to do commercial fishing is have everyone use the same gear. Have the guy with the haul seine switch to a rod and reel and participate with everybody in catching those fish.

"Please do not be unfair."

The commercial industry's Ryan Bradley told the panel, "Haul seining is hard work, and the men and women who do it provide public access to resources that belong just as much to the people in Jackson, or

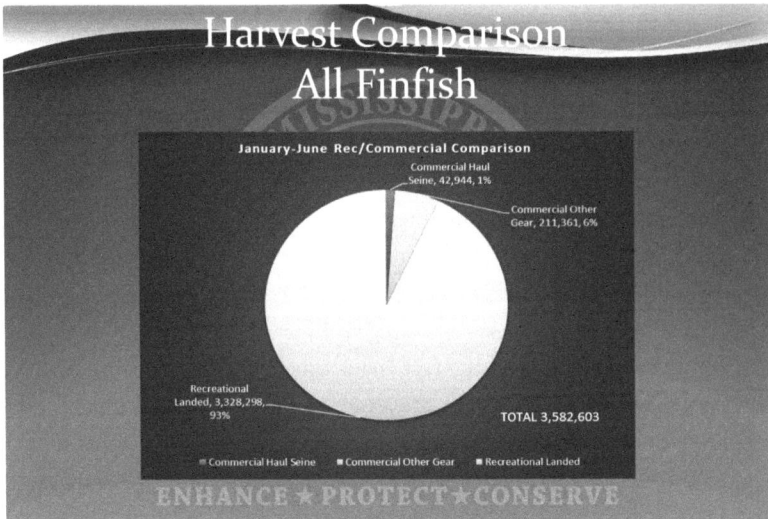

Department of Marine Resources graphic comparing sport and commercial harvests. *(Courtesy of Mississippi DMR)*

Ridgeland, or Hattiesburg, as they do to the people on this coast who are blessed to have access to the resource with a boat.

"We should be thankful for the hardworking men and women who fish commercially with these haul seines to provide this access to the public."

SEINE SCHOOL.2

Eight months after the commission's first session on the "haul seines," the controversial gear was still in use. DMR staff had used the time to collect additional data on the fishery which they reported to the panel at its May 2018 meeting.

Most of the waters that were closed to haul seines and other net fishing were within one mile of the shoreline, so the scientists did a "shoreline analysis." They found that the state's marine waters contained a little over 4.8 million linear feet of shoreline.

During the winter, net fishermen could work along just 20 percent of the total shoreline; in the summer, when popular sport-fishing spots like Cat Island were off limits, their access was reduced even further, to about 15 percent of the marine shoreline.

There was no "recreational equivalent" to the commercial closures, stated Hill. "All shorelines, waters, islands, are available to the recreational fishery."

The DMR encountered the first "haul seine" on September 11, 2015, when the agency began to track the number of people using the nets, said Hill. "We began meeting with them and looking at their gear, issuing them some certificates and approving their gear … as through the regulation, and we have tracked this throughout the years."

By 2018, 16 fishermen could legally fish with a haul seine.

The biologists also analyzed the "harvest metrics" of the top ten commercial and recreational species, which they identified as spotted seatrout, red drum, sand seatrout or "white trout," striped mullet, sheepshead, southern kingfish or "ground mullet," black drum, Atlantic croaker, southern flounder and Spanish mackerel.

From 2015, when the agency began to track the "haul seines," through 2017, sport and commercial fishermen harvested an annual average total of nearly 4.5 million pounds.

The recreational fishery accounted for almost 4.2 million pounds, or 94 percent.

The total commercial harvest averaged nearly 270,000 pounds; 225,432 pounds, or five percent of the total, was caught with commercial gear other than haul seines, including hook and line, cast nets, and gigs. The "haul seine industry" accounted for less than one percent, or 43,968 pounds.

The biologist's sport/commercial comparisons irked the CCA's F.J.

Eicke, who told the panel, "The information that Matt just provided is really a repeat of what we heard on September 19 of 2017, with some additional and equally irrelevant information as far as we are concerned."

The overriding concern of his group was that "these nets are constructed in such a way and are being used in a manner that do, in fact, catch fish by gills … and are de facto gillnets as a result," he said, "and observations by recreational charter fishermen of these nets in use substantiate that concern."

To close the loophole that allowed fishermen to use the efficient gear, CCA had developed a simple definition that "would clearly differentiate haul seines from any other fishing gear and ensure that they are utilized in a method consistent with the existing ordinance and statute," said Eicke. An alternative to not adopting the definition "would be to admit that these nets are simply gillnets by another name and apply all the regulations that apply to that gear, including the biodegradable requirement."

The commercial industry's Ryan Bradley stalled.

"Fisheries allocation is a big deal," he said. "There are big battles that are fought at the Gulf Council level and the fisheries management all across the nation: Is it 51 percent, is it 49 percent, whatever, it is going to be split commercial-recreational.

"Here we are in Mississippi, and we've got two percent commercial for speckled trout. I think there should be an uproar over that," he said.

State fishery law allowed the commission to set an annual TAC, or total allowable catch, for the sport and commercial fishermen for all species, said Bradley. "We have a TAC in the commercial sector. Why don't we have a TAC in the recreational sector?" he asked.

"This is about conservation, and we need to be looking at limiting some of that [recreational] effort."

Jeanne Lebow, an English teacher from Ocean Springs, represented herself and the Sierra Club, she said, and she wanted to limit sport and commercial fishermen to just two types of gear—cast nets and hook and line.

That's what they did in Florida, she said, and "They do not have a problem with their seafood markets, their fish markets. They have quite a few of them and they are full. You can get pretty much whatever you would like.

"They do not have haul seines, or, as I like to call them, nontraditional haul seines, that are equipped with nonbiodegradable monofilament gill-net mesh."

Catching fish by their gills was a "very indiscriminate way of fishing," she said, whereas "People who fish hook-and-line can always throw back fish. In fact, I have thrown back quite a few more speckled trout than I have ever kept."

Commissioner Jolynne Trapani, a Bay St. Louis restaurateur who represented nonprofit environmental groups on the panel, noted that haul seines hadn't been a problem for a long time but, "All of a sudden," the recreational guys were seeing the commercial guys with the nets and "now it's a problem."

Hill, the biologist, added that haul seines had been legal and in use during his entire 18-year tenure. "There were between two and five every year, before we started tracking it. ... These nets are nothing new," he said, which was of course true of the traditional-type gear.

NOT A GOOD LOOK

As long as haul seines remained undefined, fishing with the modified monofilament nets wasn't illegal.

Recreational interests were of course trying to make them so, and their efforts got a boost after several of the netters were caught underreporting their catches.

THE GILLNETS ARE BACK!

Not long after the seiners hammered the trout around Cat Island, in the spring of 2018, Alabama's marine patrol notified Mississippi that Bayou La Batre dealers were buying "increasing numbers" of spotted seatrout from Mississippi fishermen.

It wasn't illegal for fishermen to sell their products outside the state, but they had to report those sales in the state the fish were caught.

When the spring season closed, on May 31, 23,655 pounds of trout had been reported, leaving a remainder of 26,345 pounds that could be caught in the fall season. But when investigators matched the names of the fishermen who'd sold their catches in Alabama with their trip-ticket accounts in Mississippi, they found that 11 of the 16 licensed haul seiners had failed to report a total of more than 29,000 pounds of trout.

When those landings were submitted, the actual catch for the spring season exceeded the entire annual 50,000-pound quota. The fall season was canceled, and the violators were charged and fined. One netter, who was accused of neglecting to report 12,752 pounds of trout, faced a potential fine of $31,880. (Commissioners capped his penalty at $10,000.)

(Mississippi, the last state on the Gulf to institute a trip-ticket system, had recently changed its seafood sales reporting protocol: Dealers had reported the transactions until a couple months before, when the state shifted that responsibility to the fishermen themselves. The recency of the change supported the netters' claims that they hadn't turned in their tickets because they believed that the dealers would do so.)

Marine patrol eventually determined that the annual quota had been exceeded by about 4,000 pounds which was biologically insignificant, especially when compared with the recreational trout harvest which, in 2017, amounted to more than 2.5 million pounds. However, that was not the point, said Trapani, the panel's environmental representative.

"It's not really a number thing. I understand that, but they did the wrong thing, and now, other people have to be punished for it." It was a shame, she said, "for the other hardworking fishermen that go by the rules, abide by them, and someone takes their season away." (In addition to the netters, the commercial trout fishery included about 100 rod-and-reel fishermen.)

CAT FIGHT

After commissioners were briefed on the underreporting scandal, some moved to push all haul seines a mile away from Cat Island.

"This will mirror the other islands that are out there that already have this one-mile exclusion," said Mark Havard, the panel's recreational representative.

Steve Bosarge clarified that haul seines were not specifically excluded from the other islands. "What happens around the rest of the barrier islands is there is *no* commercial activity allowed," he said. "There is no law that says you can't use a haul seine."

In the mid-1990s, during the heat of the first net-ban battle, the National Park Service excluded all commercial activity, except charter fishing, from a one-mile zone around each of the major barrier islands under its jurisdiction.

Mississippi had given up its water bottom rights around those islands, but because it shared ownership of Cat Island with the federal government and some private entities, the state was able to retain control right up to its shoreline. So, the commission had the authority to establish rules and regulations for the island, including the establishment of an exclusionary zone.

Unlike the recreationally tilted National Park Service, which concocted reasons to exclude commercial nets—"an apparent reduction in available fish"—Mississippi's 1997 statutes required fishery regulations to be based on the "best available science," said Bosarge.

"What is your science?" he asked Havard.

"I do not have any scientific proof," replied Havard, "other than the eyes that are seeing what has gone on with the destroying of the grass beds and they are bringing these haul seines all the way up to the sand, on the sand. I just want to exclude haul seines."

(Recreational anglers initially couched their bid for Cat Island as a protective measure for sea grasses. When DMR scientists reported that grass beds had been declining since the 1940s around all the barrier islands *except* Cat Island, where grass beds were *expanding*, the sportsmen downplayed that ecological pretense.)

"I see it as discrimination against one class of people," said Bosarge. "We've got eighteen commercial fishermen and 82,000 recreational fishermen. This reminds me of the time when we have a certain sector that has to go to the back of the bus because the majority outweighs the minority, and that is kind of where we are now. You are going to take out the rights of 18 people."

"Just to remind you, 11 of those 18 are currently under investigation," Havard responded.

Trapani added that the discrimination wasn't against the fishermen, it was against the island itself because Cat was the only major island that wasn't "protected."

"A lot of people did not realize that this island was not protected and didn't have that barrier. I know, I got the calls, and I told them, 'I'm sorry. They're not breaking the law.' They weren't breaking the law by catching the fish. They were breaking them afterwards, and they are the ones that caused the attention on themselves."

The commission directed its staff to prepare a notice of intent to establish a one-mile haul-seine exclusionary zone around Cat Island. That initiated a regulatory process that would occupy several more meetings over the summer, prompt a public hearing in early September, and culminate in an up-or-down vote later that month.

When the panel reviewed the notice in July, Bosarge came out swinging: "We've got a 50,000-pound quota on spotted seatrout and we've got 1.7 million pounds on the recreational side. We've got every island closed except for Cat Island, and Cat Island is closed seven months out of the year. I mean, what more do you want?" he asked. "It just gets a little bit infuriating."

Havard responded, "We are not trying to eliminate commercial fishing altogether around Cat Island, just the gear type.

"That's like saying you are going to have gillnets, but you've got to have them out of biodegradable material. You just outlawed them because there is no material, and it's the same thing you are doing now."

Bosarge objected to the closure because it was politically motivated. "This is not a resource issue and, therefore, that's why I am so strongly against it," he said. "We are strictly voting on a political issue—one group that doesn't like the other group being there."

Even the commission's seafood processor Richard Gollott came out in favor of excluding the seines because, saying in August, "Recreational fishermen have had wade fishing around Cat Island for many years, and there was no reason for this bunch in Jackson County to run over there and try to raid this fishery."

Gollott wanted to push haul seines off Cat Island and have any other type of net for spotted seatrout be approved by the commission.

"What is the justification for this change in the rules?" asked Bosarge.

"Like I said, it's been traditionally a wading fishing area," Gollott replied. "Up until now, commercial people haven't gone up on Cat Island and tried to chase the fish into the nets and stuff like that.

"This was a bunch of rogues that really did this," he continued. "These people have been outlaws all their lives and, unless you do something to change it, they are going to be back. They seem to really love aggravating the recreational fishermen."

In public testimony, Tony Trapani, a past president of Mississippi's CCA, supported the one-mile exclusionary zone: "You keep on putting them nets out there and there is not going to be any reason for them to keep netting because there are not going to be any more trout. They will need to find a new job anyway. Change the nets to hook and line, and they can stand there and wade fish right next to us. ... Put a rod and reel in their hand, and let's see how many thousands of pounds they can catch."

Cecily O'Brien, of Pascagoula, who netted around Cat Island with her husband Tommy, objected to Gollott's characterization of "hoodlums" from Jackson County fishing at Cat Island. "I do personally take offense to that," she said. "I'm a full-time commercial fisherman, and I just want to point out how very hard we work as commercial fishermen.

"Just like you, we wake up and we prepare for our day, but, while you might wake up and drink a cup of coffee and maybe you read the paper before you drive somewhere to work, we wake up and we check the weather forecast, to ensure our safety on the water, make breakfast and lunch and grab our boots, our slippers and our gloves, and head out before the sun even rises.

"Our days are very long, hard, physically challenging ones. We fish in the winter's harsh cold and the summer's heat. It's a hard way to make a living, but one we love," she continued. "It is my job. It is my livelihood. It's how I pay my bills and provide for my family that is at risk here," she said.

"I'm asking you to consider all of this before making your decision about a Cat Island exclusion zone. Everybody in this room knows that it is impossible to catch a fish outside of that one-mile zone."

Gollott apologized to O'Brien, explaining, "The people who were caught going to Alabama with the fish and not turning the quota in were from Jackson County. For years the rogues in the oyster industry and everything else have been from Jackson County. I will apologize to you.

I didn't mean personally."

"Well, thank you because we do work very hard, and we do abide by the law," she said. "We don't have a ticket. We try to do everything by the book. That's just how we operate. It is unfair to categorize everybody into one group."

Haul seiner Martin Young introduced himself to the panel saying, "I'm a lifelong, lifetime, full-time commercial fisherman. ... I have relied on nets for my survival to feed my family my whole life, since I was married. This is hard for me.

"I fished them seines for sixteen to eighteen years, before any of this started. Since the ban in '98 we were not allowed to use a gillnet no more, and I never have had a cotton net, ever—that was passed, and it wasn't available.

"I was a net fisherman with no job, unemployed, and was forced to start haul-seine fishing," said Young. "What y'all are trying to do here about Cat Island, y'all have had no scientific data, y'all have had no studies done, and you are wanting to take away something that y'all have admittedly set up there and said you don't even know what it is.

"Well, I know what a haul seine is because I have built them with my hands and used them."

"As far as like this guy who just left here saying he couldn't catch a trout, well, he's got Petit Bois, Horn Island, Ship Island, Round Island, he's got all of Harrison and all of Hancock Counties, and all I've got is a little speck of Jackson County and Cat Island and y'all are wanting to take that. I have nothing left.

"I can't talk no more. Y'all are destroying my life."

One hundred and two people attended the public hearing on September 11: Of the 25 people who commented, 17 supported the exclusionary zone—"for our children and grandchildren"—and 8 opposed it. The commission was scheduled to vote the following week, on September 18.

Over the course of debates, the proposed haul-seine ban had morphed to include *all* finfish nets. So, the motion before the panel on the 18th called for eliminating the netters' five-month winter season on Cat Island and permanently prohibiting all finfish netting within one mile of its shoreline.

Before the vote, Chairman Bosarge asked for comments from supporters and opponents.

Mississippi Commercial Fisheries United, which represented 251 commercial fishermen, opposed the motion. Executive Director Ryan Bradley read a statement saying, "We currently do not believe that banning all commercial nets from Cat Island is warranted. It is in the best interest of the State of Mississippi to ensure that our fisheries management decisions are based on the best available science and not simply opinion or conjecture." Bradley requested additional scientific data on the proposal as well as an economic impact study.

CCA of Mississippi chairman F.J. Eicke supported the expanded motion, declaring that "Cat Island is a sensitive area with essential habitat characteristics that demand protection." Furthermore, Eicke said, "Many of our members have expressed concern that placement of nets close to the shorelines of Cat Island will endanger the submerged aquatic vegetation that is so prevalent near the shorelines."

Final adoption of the motion failed with a tie vote between the seafood and recreational fishing representatives. It was a break for the netters.

The swing vote on the five-member panel would normally have fallen to Jolynne Trapani—a CCA member who occupied the environmental nonprofit seat—but her term had expired in June. By September, the governor had not yet reappointed or replaced her, which, Bradley noted, was lucky for the netters.

"She's a restaurant owner and you'd have thought she'd have been on our side on this, but she wasn't," he said. Fishermen may have won the battle against an "unjust political attack with bad information coming from one side," he said, but the war wasn't over. "This is not going to be the end. It's going to come back around in one form or another. But we'll keep fighting it."

TASK FORCE

When the Commission on Marine Resources failed to ban commercial netting around Cat Island, sportsmen reflexively turned to their legislators who were soon circulating bills that aimed to ban *all* finfish netting on the entire coast.

Commissioners pursued a more nuanced approach—they wanted to take the controversial pseudo-seines out of the water yet allow the few operators of true seines to continue working.

The panel's charter-boat representative, Ronnie Daniels, started the ball rolling in November with a motion to form a task force that would identify and define the types of gear that would be legal to use in Mississippi waters.

"We agreed that a task force was a fair proposal," said the commercial industry's Bradley. "Because in all honesty, everybody knows that a gillnet with a nylon bag on it is not a haul seine. It's still a gillnet. So as much as I was enjoying allowing our guys to kind of stick it to the CCA, we were feelin' the political pressure."

Daniels wanted the working group to include Natalie Guess, the commission's newest member. Guess, of Diamondhead, was a member of the Land Trust for the Mississippi Coastal Plains and other worthy groups and had been appointed to the environmental nonprofit seat in October. She would be the least-biased commissioner on the panel, said Daniels, "and I think she would do a good job."

Commissioner Gollott also recommended that Bosarge sit on the task force, "because he knew about nets." The DMR's executive director, Joe Spraggins, and agency staff would choose the other members.

NO TIME TO WASTE

Before the start of the 2019 legislative session, the commission held a forum "where our legislators came in and we gave them our wishes and they gave us theirs," reported Gollott at the panel's December meeting. "The one thing they made very clear is they wanted the commercial nets, finfish nets, off of Cat Island," he said.

"They told us flat out if we didn't do something, they were going to do it, and the last thing this commission should want is the Legislature to get involved in this."

To forestall that, Gollott wanted the commission to ban haul seines around Cat Island even before the task force met. Sacrificing the gear around the island could allow the seiners' only other productive site to remain open, he said.

"I'm not against Martin or any fishermen. What I'm thinking is if the Legislature gets ahold of it, they are going to ban all nets, and some of these people are fishing in Jackson County with these nets, and they will still be able to fish in Jackson County, if we just put this ban around Cat Island. That's my way of thinking. It's not that I'm trying to hurt anybody."

Seiner Martin Young didn't like the idea. The little bit of fishable area in Jackson County was too small and overrun with recreational boats that scared the fish off, said Young.

What's more, he had "as much right as any citizen to fish" around Cat Island. "I fished that island when I was ten years old with a net with old men that are dead and buried. ... Mr. Gollott says Cat's long been a recreational island with wade fishermen. Well, I have fished my whole life for my living, and I did catch redfish at Cat Island. I caught sheepshead at Cat Island. I caught drum at Cat Island, and I caught mullet at Cat Island. I picked hardhead catfish until my fingers bled, out of gillnets at Cat Island. No, it's not just a recreational island," said Young, who implored the panel to not take it away. "I'm settin' here, all but on my knees, beggin' you."

Gollott withdrew his motion but urged the panel to act: "The legislators gave us an ultimatum to do it, and I think that we need to move forward with it."

DMR executive director Spraggins reported that the makeup of the gear task force had been determined and would include two commercial fishermen, two recreational fishermen, two DMR scientists, one marine patrol representative, two scientists from the University of Southern Mississippi and Mississippi State University, a NOAA representative, a gear specialist, and the two commissioners. Neither the commissioners nor agency staff would be voting members.

In January, the commission moved to place a temporary moratorium on all netting around Cat Island. It was to take effect February 1 and remain in place until the task force completed its work.

The seafood industry's Bradley objected to the motion. The agency had already revoked the tags for the other nets so there were just two or three fishermen using the traditional haul seines. "It's a heavy-mesh seine, small mesh, does not entangle fish, very difficult to use," he said. "So, I just don't see why we have to do this for two or three people. Can we not live with two or three people out there fishing?"

Commissioner Daniels responded that it wasn't those "two or three people" that concerned him. "It's the unknown that we are worried about," presumably the introduction of another unforeseen contrivance for catching fish in commercial quantities.

Haul seines were made of coarse nylon twine and could weigh over a ton. Unlike monofilament gillnets, which "entangle" fish as they try to pass through their meshes, seines "entrap" fish by surrounding them with webbing. Word games aside, the gillnets are more selective than either seines or the hooks of recreational anglers. *(Ryan Bradley photo)*

WHAT'S A HAUL SEINE?

The Gear Type Task Force adopted a definition for the haul seine at its first meeting on January 22, 2019.

Two days later, the commission met in a special session, unanimously accepted the definition and passed the Notices of Intent to start the regulatory process that would incorporate it into the state's fishery regulations. On the same day, DMR staff filed the Notices with the Secretary of State and posted them on the agency's web page. A legal notice with the definition appeared in the *Sun Herald* on January 26, and on February 14 it was aired at an uneventful public hearing.

The task force defined a haul seine as "A net set vertically in the water column and pulled by hand or by power to capture fish by entrapment through encircling and confining fish within itself, the shore, or bank, as a result of mesh size and construction. Haul seines shall be constructed of a mesh size not to exceed fifteen-sixteenths ($^{15}\!/_{16}$) inches bar and one and seven-eighths ($1\!^{7}\!/_{8}$) inches stretch and shall not be constructed of monofilament."

The commissioners allowed nets conforming to the haul-seine definition to be deployed right up to Cat Island's shoreline during the winter commercial season. Gillnets rigged as haul seines were history.

"Essentially, they made it clear that you can't use the monofilament net with a nylon bag on it and call it a haul seine. So, all those guys who were using those nets were out again," said Bradley. "It alienated them because they were enjoying what they were doing. But the reality is that a lot of [the monofilament-net fishermen] that were doing it were bringing flak on the industry. They were posting on social media of them running these nets. They were not reporting their trip tickets on time. So, they'd really brought it on themselves.

"At that point, we said, 'Well, about all we can really do is protect the people who've been using the real gear for decades.'"

POOR BEHAVIOR

An old regulation passed to appease recreational interests closed Cat Island to commercial netting from May 15 to September 15. The winter season, when the island was open to both sport and commercial fishing, typically lasted into May but this year the commission closed it February 1, while the task force completed its mission.

After 19 days, the panel adopted the task force's definition and lifted the moratorium. So, Martin Young and the O'Briens teamed up and went back to seining around the island.

Then, on April 16, the DMR's marine patrol received a number of complaints about a netter on Cat Island. Enforcement sent a unit out to the area and made contact with the boat that met the description given. The agents checked Young. "They seen we was perfectly legal," he recounted.

Later the same day, the seiners were charged by a recreational fishing boat whose driver, said Young, attempted to damage his net. The Bay St. Louis angler, a former president of the CCA's state chapter and "an icon of the tourism industry," then threatened to cut the seine with his knife. Meanwhile, he unleashed a barrage of profanity against the netters, and while this was ongoing, Young called the DMR. When the marine patrol agent on the phone overheard the sportsman calling the fishermen everything under the sun he sent a unit to the scene.

The sportsmen were frustrated. They'd gotten so excited at the prospect of winning Cat Island that they'd worked themselves into a frenzy at sport get-togethers and on social media. Yet they'd been thwarted by the commission, and now they were being outmaneuvered in the Legislature.

2019 LEGISLATURE

In January, commercial lobbyist Ryan Bradley had asked the commission, "Can we not live with two or three people out there net fishing?"

The sport fishermen's answer was "Hell no!"

Cat Island was eight miles long, yet nowhere near big enough to share with the seiners, even for just part of the year.

When the 2019 Legislature convened in January, a bill in the House of Representatives called for the prohibition of haul seines within one-half mile of any shoreline on the coast. House Bill 98 was later modified to exclude haul seines from Cat Island only.

In February, the bill passed out of the House almost unanimously, then moved to the Senate "and that's where we had an ace in the hole," recounted Bradley.

Thomas Arlin Gollott, from Biloxi, was the longest serving state legislator in Mississippi history; he was first elected to the House in 1968, and then to the Senate in 1979.

Gollott had almost singlehandedly established the Department of Marine Resources and its guiding commission in 1994. He was now chairman of the Legislature's Ports and Marine Resources Committee.

"And so I go up to Jackson and I go to him and I plead my case. And he really wasn't for me in this situation," Bradley recalled. Gollott was a seafood industry supporter "but on this particular issue I could tell he was getting a lot of pressure because he would come back at me, 'Oh, they need to do this.'"

In a March 1 letter to Gollott and other members of the Ports & Marine Resources Committee, CCA leaders Tommy Elkins and F.J. Eicke asked the senators to support their bill:

"For years, the issue of improper and deleterious use of nets has been a concern in the marine waters of coastal Mississippi. HB 98 would protect and conserve our marine resources from the use of nets that gill and entangle and, more specifically, the use of haul seines for the taking of fish within one-half mile of Cat Island."

The CCA didn't "seek to eliminate commercial endeavors but rather to address issues that make the pursuit of commercial fishing adhere to 'reasonable constraints'"

HB 98 was a step in the right direction, wrote CCA's leaders, "in addressing what has become a concern for not only recreational anglers but also to commercial fishermen using more sustainable gear," an apparent reference to some commercial rod-and-reel trout fishermen who competed with the netters for that species.

In his own words, Bradley described how the CCA's net-ban bill met its demise:

"I pleaded my case to Senator Gollott, and … he promised me, when I walked out of his office one day, 'It's dead. I'm not gonna do it. Don't worry about it.'

"Well, I didn't stop worrying about it, I stayed up there in the Legislature almost the whole time until it was for-sure dead. The long story short, when it came up for a vote in the committee, we worked the committee members and were able to get them to kill it.

"Well, one of the legislators did a motion to reconsider and held it. They was tryin' to force it back on. So Gollott just set there and said, 'No, we ain't bringin' it out.' So he saved us there.

"Well, it wasn't over yet. There was another bill that was actually one of Gollott's own bills—an oyster bill. They were trying to do some oyster reef privatization to allow oyster leasing. Gollott sponsored this bill, introduced it and it passed the Senate unanimous, then went to the House.

"The guy in the House that was pushing the net ban for the CCA slipped the Cat Island thing into Tommy Gollott's bill and sent it back to him. The House voted on it and they sent it back to the Senate to vote on it one last time. I went to Gollott and said, 'What the heck?'

"He said, 'Don't worry about it, we ain't doin' this bill.' So, he killed his own bill to kill all this completely. And there was a big push, they wanted to send it to conference and work out the differences and all this. And he said, 'No, it's too dangerous. We don't want it to go nowhere.' And so Gollott just stood there, and every day it would come up on the calendar he would pass over it.

He would say, 'Oh, I don't want this today.'

"Until the final deadline for action hit and it was completely dead.

"That was another major victory for us and it really kicked CCA in the nuts because they had pushed so hard for that Cat Island net ban. They tried so many sneaky maneuvers to get it done.

"So, at this point we done beat 'em at the commission, we done beat 'em in the Legislature. We're fighting hard to keep this, you know?"

2020 LEGISLATURE

The Mississippi Gear Type Task Force described itself as "a duly constituted assembly, created by the Mississippi Commission on Marine Resources to serve the Great State of Mississippi." The group had been charged with establishing a definition for the haul seine, which was adopted by the CMR and amended into the Administrative Code Title 22, Parts 5 and 8.

The loophole that allowed gillnets to operate as haul seines was closed. But in 2019 Gollott retired, and the following year the CCA returned to the Legislature with a pair of identical bills that aimed to force haul seines at least half a mile off Cat Island. This time there would be no "ace in the hole" for the commercial sector. Gollott's replacement was one of the prime movers behind CCA's net ban.

Biloxi's Scott Delano, a commercial real estate developer, had spent about a decade in the House before transitioning to the Senate in a special election. With his legislative experience, he was well positioned when he assumed office in January 2020.

"So, we know we're probably going to get hammered," said Bradley. It didn't help that the new senator had won his seat after a rancorous battle with an industry-backed candidate.

Dixie Newman, a Biloxi city councilwoman, "raised a ton of money" and, initially, won the election by a single vote, recalled Bradley. Stunned, Delano demanded a recount and lost again. He finally prevailed in a special re-election.

CCA's "golden boy" wasn't the industry's only problem. Coastal legislators generally sided with recreational interests as a matter of course, Bradley explained. "They view the recreational fishermen as votes—many more votes than commercial fishermen and that's the way they look at it, plain and simple."

Rep. Timmy Ladner from coastal District 93, and Sen. Mike Thompson, from coastal District 48, introduced the CCA's bills simultaneously in their respective chambers. On March 10, the Senate passed SB 2720 with a 50-2 vote and on the following day the House passed the anglers' HB 561 by a 118-3 vote.

The bills then moved to the opposite chambers but before they could win approval the COVID-19 pandemic intervened. To reduce the spread of infection, the Mississippi State Legislature suspended its 2020 session from March 18 through May 7.

Fishermen made the most of the break. They reconvened the gear task force, which voted unanimously to oppose the sportsmen's bills with a resolution stating in part that the proposed laws were "not based on the best science available" and didn't address a "specific management need pertaining to sustainable fisheries."

The resolution concluded with an appeal to "Mississippi's esteemed and honorable elected officials for help," saying the two bills would "unfairly and inequitably restrict the rights and equal access of the commercial fishing industry."

Lawmakers were unmoved.

So the task force, which served only in an advisory capacity, upped the ante by asking the commission to adopt its resolution as its own.

Task force chairman Frank Parker, a commercial fisherman, told the panel at its March 31 meeting, which was conducted via Zoom, "We feel it is important that we maintain our local ability to commercially harvest seafood using traditional methods at traditional harvesting locations as a matter of national food security."

Seiner Martin Young reminded the commission that it had been "appointed to handle seafood law on the coast in the early nineties. So let the voice of the commission speak loud to uphold" the task force's position.

The bills were based upon "the greed of the recreational sector and are not fair and equitable to the commercial net fishermen," haul seiner Tommy O'Brien told the commissioners. "Once again, it will take away from us the most lucrative fishing grounds we have."

Mrs. O'Brien reminded the panel that money was tight during the pandemic. "We need all the help we can get. Any input that you can have in Jackson to help quash these bills will be greatly appreciated."

Commissioner Bosarge coaxed his fellow panelists to support the task force. He felt it "strange," he said, that there was "a group in Jackson that feels that they know more than this group."

DMR executive director Joe Spraggins, who'd previously served as a Brigadier General in the U.S. Air Force, apprised the panel that he was trying to convince legislators to stand down but was having little success. "I asked them to please allow DMR to run the operation. But it seems right now that we have a couple of new legislators" who were "getting a lot of complaints from their constituents, and the whole thing is about Cat Island."

On April 7, the commission called a special session via Zoom to finalize the task force resolution, which appealed to the House and Senate to rescind their bills regarding the prohibition of haul seines off Cat

Island and allow the commission to manage the issue through the powers it had been granted by the Legislature.

Recreational representatives Ronnie Daniels of Pass Christian and Mark Havard of Vancleave opposed the resolution. The gear task force was established to define gear types, "not to determine whether an area is permitted to be fished, or not," said Havard.

Mississippi CCA chairman Tommy Elkins, in his public comments, said that it was inappropriate for the commission to object to the bills because the panel wasn't authorized "to attempt to influence the legislative process."

Bosarge stated that in the ten years he'd been a commissioner he couldn't recall a single time when the Legislature had made changes in the panel's laws "that we did not request." The task force had approved a gear type for Cat Island and "now we've got a body that has no science behind it—it is strictly political—wanting to undo what they have done," he said.

The panel voted 3-2 to adopt the wording of the gear task force's resolution. The swing vote was provided by the commission's environmental representative Guess. "She's a woman that has a conscience to get past the male ego involved," Bradley commented.

After the vote, he told the Pass Christian-based *Gazebo Gazette*, "Today's Commission decision is a reflection of ever-changing fisheries management needs and reminds Coast legislators that commercial fishing is considered an essential infrastructure. It is critical to our nation's ability to produce our food supply domestically."

As the pandemic disrupted supply chains, the government had classified commercial fishing as an "essential infrastructure."

"That gave us a new argument that we never realized we ever had," explained Bradley. "So we appealed to the legislators and said, 'Hey, this is necessary for our food supply. It's important to our national security to produce our food supply domestically.'

"I know we're starting to take this a little bit far but that's where we're at now," he added. "It's normally a hard argument to make but now, during these times that we're in, it's not so hard anymore."

Fishermen hoped that supply chain disruptions during the pandemic would sober legislators and discourage them from supporting CCA's anti-commercial bills when the session reconvened.

They didn't.

PAYBACK

The politicians were furious that the commission had challenged their authority and assured the panel that they'd now be *certain* to pass a net ban.

In its final form, Sen. Mike Thompson's SB 2720 prohibited the use of haul seines within one-half mile of Cat Island's shoreline. His bill also prohibited the use of either a "gillnet, trammel net, entanglement net, or *like-contrivances* for the taking of finfish" within one-half mile of any shoreline statewide.

The catch-all "like-contrivances" was defined as "any net that is similar in form, function, purpose or use to a gillnet, trammel net or entanglement net," and was intended to minimize the possibility that fishermen would find another novel way to catch fish in volume.

Angry legislators—as their predecessors had during the 1990s net fights—threatened to disband the commission altogether. Although they stopped short of that, they demoted the panel by stripping it of its regulatory authority.

At a briefing after the session, DMR director Spraggins explained to the commissioners that the legislators took the letter that they'd sent to the Legislature, opposing CCA's bills, "in their mind and ... said,

'This commission just told us we didn't have the authority to tell them what we think ought to happen.'

"We told them that we felt like the commission and the department had this headed in the correct direction and that things were working. 'Why won't you just let us do this?' we asked. But it didn't matter what we told them. It was going to happen."

Bosarge responded, "So some senators didn't like that we made a decision based on science with the best facts and it wasn't the decision that they thought needed to be made politically, not on science but politically. So that was the reason they took the commission and made us advisory?"

"Undoubtedly," answered Spraggins.

Legislators meted out additional punishments for the commission's seafood representatives.

The panel's members were normally appointed by the governor and confirmed by the senate. Yet when the terms of seafood processor Gollott and commercial fisherman Bosarge expired in March 2021, the Senate Ports and Marine Resources committee blocked their reappointments.

"This is how they pay us back for fighting them on the net ban," said Bradley. "These are the same senators that led the net ban, and they don't like any of us. It really speaks to the lack of respect and disregard for our hard-working commercial fishing families."

After several months, the Senate allowed the seafood processor's slot to be filled but the panel's commercial fishing slot remained vacant for more than three years. In July 2024, a restauranteur and off-bottom oyster farmer was appointed to the position.

MISSISSIPPI FISHERFOLK
IN THEIR OWN WORDS

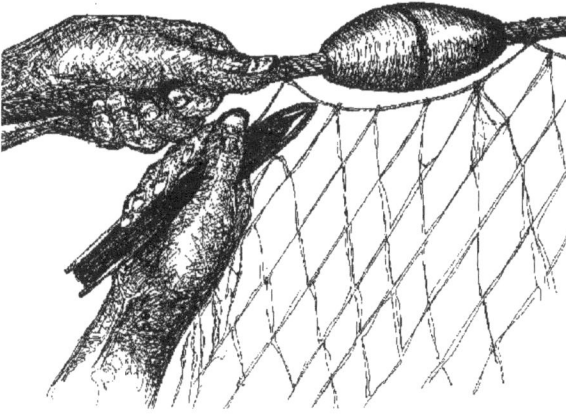

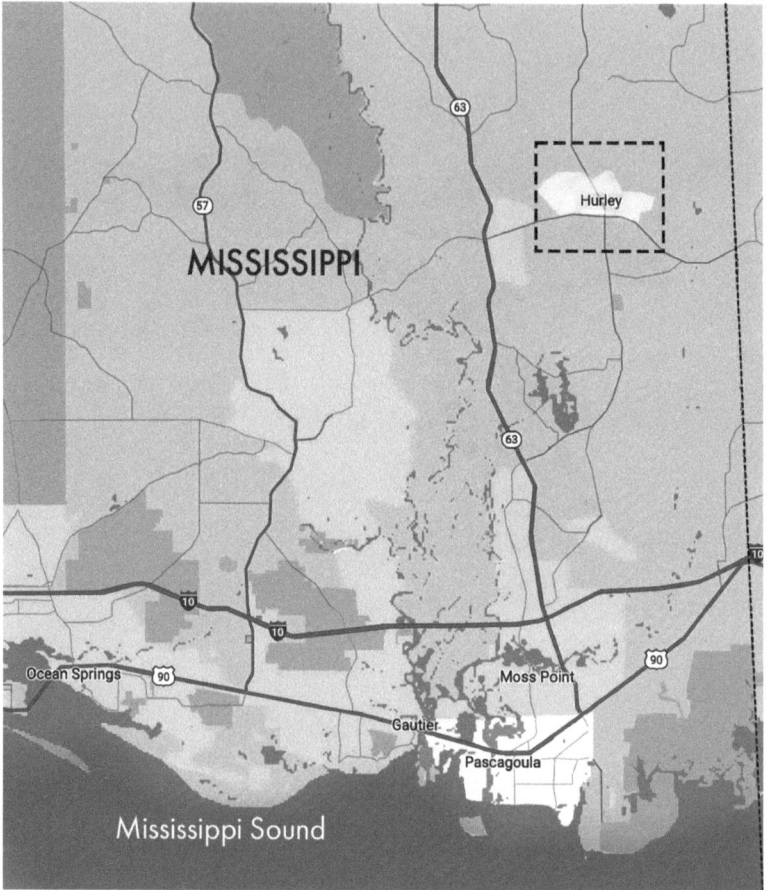

MISSISSIPPI FISHERFOLK

IN THEIR OWN WORDS

MARTIN YOUNG
HURLEY

Martin Young. *(Ryan Bradley photo)*

MARTIN YOUNG HURLEY

Commercial fishing lobbyist Ryan Bradley introduced me to Martin Young during the 2020 tussle over haul seines. According to Bradley, Young was "salt of the earth, hardworking, a big guy. I mean, pullin' in these nets all day, he ain't no scrawny fella."

Monofilament gillnets are light in weight, efficient, can either be set out to fish passively, like a spider's web, or used to encircle a school of fish. Haul seines are set around a school of fish, or an area that's likely to contain fish, then dragged down and hauled back aboard with a winch. They're of heavy-duty construction, complicated to piece together, and not for amateurs.

So, when Mississippi's Commission on Marine Resources forced fishermen to work with unworkable nets in the 1990s, there wasn't exactly a stampede into the fishery.

Martin, who'd spent a lifetime netting fish, knew how to build a seine. So, he and his brother Rufus hung one and spent the next couple decades working it, mostly around Cat Island, where they caught the usual complement of the gulf's coastal finfish.

Then, in around 2015, some resourceful fishermen began to exploit a regulatory loophole that enabled them to pass off their monofilament gillnets as "haul seines." Environmentally, their nets were perfectly sustainable, but politically they were radioactive.

When sportsmen caught wind of their activities around popular Cat Island, Martin was caught in the crossfire.

In a classic case of throwing the baby out with the bathwater, the anglers tried to ban all netting within one mile of the island. Fishermen beat that 2018 effort in the commission. Then, in the 2019 Legislature, the sportsmen tried to outlaw haul seining within one-half mile of Cat Island. After the fishermen beat them again, the anglers came back harder the following year.

This time, they steamrolled the seiners from the get-go but before CCA's politicians could close the deal the COVID-19 pandemic shut down the 2020 Legislature. After a nearly two-month lull, the session was about to reconvene in early May when I first called Martin to get his take:

2020

I'll be 57 in November. And that's a bunch o' hard ones!

I had two girls, no boys. And my grandbabies are four and two. My brother, same thing, he only had two girls. He's 61 and now he's on the pogy boats. He's in Empire [Louisiana] working at Daybrook Fisheries.

Was my dad a fisherman? Yessir. We run a family business that we lost because it was a wholesale, mostly mullet, sheephead, them-types fish. We didn't do a big retail market. We were more of just a wholesale truck-business market instead of like a showcase market.

That business was 32 years old, and we lost that when they took them gillnets away. And me and my brother went to work for Clark Seafood. We had to go to somebody that handled a volume of fish instead of just a few.

I was born and raised in Orange Grove, down there around Bayou Cumbest area. And I got wiped out twice in seven years. Hurricane George took me out, me and my wife lost everything and we was a-comin' back when Katrina hit and we lost everything again.

248

MARTIN YOUNG HURLEY

Hurley, where I live now, it's almost an hour drive for me to go down there. I'm gettin' older, but I'm lookin' at my boat settin' up here right now. That's what I'm still plannin' to do, if we survive this Cat Island deal.

My boat's right around 24-foot, aluminum. I got a 150 Yamaha on it. It's fully decked in to repel water off and for a poor little fisherman, I reckon it's a pretty nice boat. It ain't the best but it'll do.

We trailer our boats down there and launch at Gulfport or whatever. We still fish out there at Cat Island—it's the only island we got. I mean National Seashore took Petit Bois, Horn, and Ship from us. So that was the only one we had left and the only reason, they said, was because private owners and the state bought it. It wasn't federal. The rest of 'em are federal-bought-up islands.

If they take Cat Island, all that's left is one little place. And there are so many recreational fishermen and all there nowadays. You know how a spot is that gets run over eight or ten times a day? There's nothin' there. And that's what they're wantin' to leave us with.

So, they're doin' like they done back in the Civil War, they're just gonna starve us out. If they can't put us out of the way, they'll just starve us out. That way they can say, "Oh, you can still go. We didn't stop fishin'."

But they know they stopped you because you're so limited that you can't do nothin'.

I've done heard that so many times since in the 1990s, I've got that down: "We're not stoppin' you, you can go fishin'," but you can only go here, with this, in this little spot for this little bit o' time, and on and on and on, whatever they can come up with to hurt you.

What do I catch? Well, they've about got it to where the trout and redfish, they keep that closed more than it's open. But we fish for 'em. And then sheephead and drum. And then mullet. But mullet sales don't seem to hold up too good for us no more, like it used to.

The roe run? Yeah, we fish the roe a little. But mullet jump. If you set 'em with a gillnet you know how they like to jump? Multiply that by about a hundred and you can imagine what it's like tryin' to catch 'em in a haul seine.

You'll start out thinkin' you got 5,000 pounds and you end up with 200.

My net is 1,200 feet long and the mesh is from 1½ inches to 1¾ inches, somethin' like that. The smaller the mesh, actually, the less the trash fish that gills off. And the better it is on your undersized fish that you're gonna pick through and throw back.

Me and that brother of mine was the only two that used seines for all them years since the ban. And we wasn't so much on the publicity and the public eye. We was never about that. We was all about our work and doin' our fishin'. Tryin' to stay sustainable, not killing off all your fish too small. I mean, we worked with all that but we wasn't in the public picture. Nobody really knew what we were doin' until they started that deal three years ago where they done these tags on them nets and was callin' 'em haul seines. And they was just gillnets with a little piece o' nylon on the end. They knew it was a sham.

In twenty years, we never had a ticket, never did nothin' wrong, and there wasn't a lot of complainin'. But as soon as they put them monofilaments back and called 'em haul seines, that's when people went to callin' and complainin'. And we didn't last a year. Or almost two.

Did we have to get an exemption to use our seines? No. They had no regulations, no rules, no definitions, nothin' regulatin' a haul seine until this past year.

Ryan nominated me to be on the gear task force and we made a definition of the material type, the mesh size, the length. We done all that and we no sooner got that done, here it is 3 or 4 months later and they're up here in Jackson trying to pass this bill.

So it's kind of like a double standard: We're over here at the commis-

Sorting the catch. Martin removes "puppy" black drum from dip net on a January 2, 2019, trip. After deckhand Tommy O'Brien Jr. dips catch from bag, marketable fish are saved, undersized ones are returned to the water live. Brown pelicans outside net wait for discards. *(Ryan Bradley photo)*

sion trying to fix it all and voting with them and agreeing, and then they're up there in Jackson—the senators with all the recreational people—trying to slip bills through to put us out.

It's kind of an underhanded deal right there. But the commission, Kirk Fordice was governor in the '90s, and he appointed that commission to handle seafood laws down here on the coast and that's what they was appointed for to do. But now the Senate and the House are trying to pick up the sportfishermen's bills to favor the recreational people. That's what it all amounts to.

We're outnumbered so bad; I mean you could say three or four of us to thousands. That hurts. But we ought to be considered historical fishermen anyway because we done it all our lives. We was raised in fishin' families, we done the haul seinin' when nobody else would.

Them people over there, they don't know what it is to use that seine like me and that brother of mine did, workin' everything by hand. They don't understand the physical manual labor that's involved in that. They ain't got no pity.

I stood up there and just spilled my guts and just bled my heart out to 'em and told 'em, "If you can't do nothin' else, treat me fair, for once, and just buy me out."

In Florida they paid 'em for their nets, job trainin', and helped 'em out. In Lou'siana they gave 'em different options. We got nothin'. Mississippi, when they done this monofilament net ban, in '96 I think it was, we didn't get nothin'. All we got was "Goodbye!"

I could go on and on, man, about how this thing has went because I have had to live it. I mean, it has been my life. I've stood up at them meetin's and cried like a baby tryin' to save what I knowed I'd worked over 20 years at, honest and legal, and they was takin' away because of the crime that had went on callin' the mono a haul seine.

I was born and raised in a skiff and I depended on it a hundred percent my whole life. A lot of these other guys were part-timers. That's a big difference when it comes down to the bottom line.

They didn't want to strike fish and find the fish first, they wanted to go home and make their money in bed. They went to runnin' 'em straight out, just like they used to. And the sad part about it, it was the same ones that was doin' all the outlawin' and causin' the problems back in the '90s that all went back to fishin' as soon as they heard they could get a mono tag.

It was pretty ridiculous the way it went. Nobody else done it. It was me and my brother and then the O'Briens, after about five years back they built them a seine and they started tryin' it. And we was pretty good friends with 'em so they started workin' with me and my brother. They done it on their own a couple years and was usin' it like an old-style haul seine, pullin' it to the hill [shoreline]. Then they started fishin' with us and we started showin' 'em how to make 'em work where you didn't have to come off the hill. That's old school.

(Illustration in Appendix shows traditional method of hauling seine to shoreline. Peter Matthiessen's 1986 book, "Men's Lives," describes the technique as practiced by Long Island haul seiners. They harvested striped bass by launching their dories into the surf of the Atlantic Ocean and hauled their net back to the beach with winches or trucks.)

Me and that brother of mine fixed it so that we could haul off the hill. It'd be hard to tell you how we done it, but the way we had to cut certain tapers in it, the way we built our ends for our bag and all, we had to change all that. The way we hooked it to the boat, and the big part of it, the way you pick it all up at the end of the set is where you catch your fish. And we didn't share that with a lot of people because we was in fear of 'em a-gettin' it stopped, for over 20 years.

But, yeah, we will still drag a haul up on the hill if we catch the fish in the right place. We still do it both ways so, yeah, puttin' us a mile, a half mile or whatever, off the hill, that's devastatin' to us, for sure.

Cat Island? Yes, we do need Cat Island. That's the only gulf fishing we've got left with good access to redfish and drum, sheephead and speckled trout. And it's basically bein' took away from us by the recreational people who know these senators and representatives and all that's tryin' to have it done over the commission and what has been decided down here and voted on by all of their specialists.

On that gear task force, we had NOAA guys, state guys, we had law enforcement, we had recreational fishermen and commercial fishermen,

COASTAL EROSION PART TWO

all sat on that task force. And everybody agreed on everything that was
done to establish a definition for haul seines. It was all unanimous votes
on everything we done. And then the commission approved it. And now
they're tryin' to change it in Jackson. That's the best I can explain it to you.

Did we specify a twine size? No, it was material in general, all it's got
to be is nylon. And the mesh, it can't be nothin' larger than a certain size.
The smaller you get ain't never gonna hurt. The bigger is what gets 'em
by the gills. We all agreed on, I think, an inch-and-three-quarters, for as
big a mesh size as you could go.

What's gonna happen? Well, all I know is what Ryan had told me
when he was up there. Myself, I hadn't been to Jackson. Now, as far as
Biloxi goes, I've been to meetin' after meetin' after meetin.' I went to task
force meetin's, on this gear task force. I mean I've tried to be a little more
involved but as far as saying now what I think's gonna happen, I don't
know but they might act like your friend at those meetin's at the DMR
building, over at that commission, when they actually ain't your friend,
either. So I really couldn't tell you what's gonna happen.

The most I can tell you about the Cat Island thing is they're tryin' to
politic things to take it from us. And your normal little group of
recreationals [CCA] is callin' these senators and representatives that's
causin' the problems, like normal. In Jackson, we're totally outnumbered.
We can't beat that, I don't feel like we can. And the recreational people,
more than likely, are gonna win the vote up there.

So we put a resolution in to try and get the commission to send a
resolution in to stop that bill. Because what the task force had done and
voted on was supposed to be made law.

At the gear task force meeting, Ryan told me what I needed to do, he
coached me through it because, like I said, I am not a political person.
But I motioned for the resolution, alright? They said they couldn't vote

on our resolution, they had to make their own resolution from the commission before they could vote on it.

And it passed. After that, as far as I know, the Corona thing hit, and I don't know if the resolution they sent to Jackson done any good or not. But I don't think nobody knows nothin' about that yet.

Why have a commission if they're gonna oversee what they're doin'? They're supposed to be handling seafood law down here on the coast, they ain't worried about bream and bass and freshwater fishin'. This is all supposed to be saltwater seafood stuff that they're lookin' at.

It's always been the same battle. It always comes down to votes and money. It shouldn't be that way but the words "fair" don't have no meaning, doin' what's right, that's got no meaning.

That's why I asked for 'em to consider us as historical fishermen and leave us alone until our time is gone. But they won't do nothin' that might benefit us. No matter what you ask for, they won't agree to nothin' that would benefit you.

We asked for 'em to do a moratorium where we could sell our license when we got ready to retire. Limited entry. They wouldn't go for limited entry. We've asked for a bunch of stuff, but they wouldn't agree to none of it. Like I said, I got up there, man, and just cried like a baby. I mean it's really hard, it really affects you, it's affecting your life, but they don't care. They don't care a bit.

2024

[When I interviewed Martin in 2020, he predicted that legislators that year would give sportsmen their Cat Island haul-seine ban. They did.

So, a few years later, in September 2024, I checked back with him to see how he was faring. The following is an edited transcript of that conversation:]

Did they push us a half mile off Cat Island? They absolutely did. They got that Ladner and that Delano elected into the Legislature and they introduced them bills about Cat Island the day they went in. They was the first bills, I think, they introduced.

Talk about puttin' a knife in our backs, that did it when they took Cat Island because it was the only one we had left, a piece of it anyway. And that was only for six months out of the year.

But, like right now, I could be back fishin'. I think it used to open back up like August 25. So we'd be goin' fishin' now. We could go catch some sheephead and drum, redfish, trout, whatever's there to catch right now. At least we'd have some areas to go fish. But we ain't got it no more.

They got their way and they done it in a sneaky dirty way, if you ask me. Anybody that was a lifetime-dependent, commercial man, you basically just got pushed back, pushed back, pushed back to where there's nothin' left.

Where can we work now? There's not much. Think about it—all of Harrison and Hancock County beaches is all jet-ski rentals, just tourism in general that took all that. The only thing that we did fish off of them two counties was Cat Island.

Like we talked about before, we don't have to set our net right on the shore. But we can't do nothin' half a mile off. Once you pass a half mile you go over what we all call the "drop off."

From the shoreline you go out to your inner bars and then you drop

256

off further and then you really drop off. It gets deep. The only thing we could fish out there, if we could drift something like we used to with the gillnets, you could drift fish mackerel, stuff like that. Ladyfish, bluefish.

But as far as catchin' your trout and redfish, sheephead and drum, all that stuff's pretty much gonna run from the shoreline to the inner bars. They'll fall off on the edges into the deep water but that's not nowhere that a smaller boat like we work on would be able to fish.

Why was it so important for them to have Cat? Because the people right there with access to that place, at launches over there at Pass Christian, Gulfport, all them little boat launches that's on that beach over there, they all got nowhere else to go. They either got to go up Back Bay Biloxi, Back Bay St. Louis or straight out to Cat Island.

They fought it tooth and nail, they hated nets. Them people in that part of the world hated nets, they don't care, a net's a net is a net to them. It don't matter what kind of net it is.

They think you catch 'em every day because you got a net. They don't understand that. It's just like you don't catch shrimp every day just because you dropped those boards and net off the stern. They think you can. That's exactly how they look at it—the commercial guy's got a net, he's gonna kill 'em. The poor little fish.

I'll tell you, the recreational has exploded so bad. It's like we are in the mainstream tourist capital now. That's what they campaigned to do, to basically turn Mississippi, this coast here, into a tourist destination and that is what's happened. That's what they wanted, and they got it.

I haven't been on the net boat since June. You go down there and you can't hardly get on the parking lot to launch your boat, on weekends especially. You can't even get to a boat ramp on the weekends, there's so many of 'em.

Man, it's just a different world than what I seen twenty years ago.

We'd go, just twenty years ago, and you might see a dozen people where now it's 200, 300! It's unbelievable.

And everything stays so disturbed and occupied anymore, man, it is hard to make ends meet anymore with what we got left.

See, let's take and say you got three areas to fish. And you go fish all three areas every day, they don't never get a break. Well, you're gonna quit catchin'. You can't keep going to the same area over and over and over again, day after day. It just don't work that way. It needs time to build up, to rest.

But, in between, you think it's restin' but you got a thousand crab pots everywhere, they're comin' in there and doin' their round and round, disturbing everything every day. Then you got a thousand recreational fishermen coming back and forth, back and forth, doin' their disturbin' every day. And then nowadays you got jet skis, you got people who's goin' out in boats that ain't even got no business goin' in the area. I mean they're in the area just to go boat ridin'.

There's just so much that goes on in the area that, in my opinion of it, the fish has changed up where they can and can't do what fish do anymore. I think they have to stay in the deep water more. They have to stay in a river, up under a dock or up under a skiff or a barge. I mean they have to spend more time in them kinds of areas because everywhere you can fish, they disturb much of the time.

I've been doing a lot more rod and reel than I have with the net. It got to where the hardheads—we had a bumper crop this year of catfish, trash fish. And I quit working maybe in the middle of June. And then we rod-and-reel fished.

I've got a boy fishin' with me and we rod-and-reel fished right up till about August. And then it got to where we had quit catchin' enough to make ends meet. We was catchin' a few trout and redfish. And that slowed down when that heat set in and I ain't really done nothin' the whole month of August.

We mostly rod and reel for trout. I've kept my hook-and-line license for a long time. But you're restricted with that too.

We can't go fishin' in the bays north of the CSX railroad. Everywhere the fish migrate to and from we can't go fish. But all the charter boats can fish it and all the recreationals can fish it. Same thing for the islands. We can't go fish the islands either. See, if they check you, you can't have but a sport limit of 15 trout.

So, I mean, even on that aspect, it still ain't fair—even on the commercial hook and line. They let the charter boats make their livings out there, but we can't go out there and fish.

If we could have what they kept from us out there on those federal islands you still might could survive and make at least a get-by decent living. But we ain't had that for so many years I barely even think about it anymore.

What do we use for trout bait? We mainly fish for them bigger trout in the spring and summer and we mainly fish with live croakers. We catch them with our hook and line. We would buy these little packs of froze shrimp, cut about five pieces out of a shrimp and then use a small bream hook. You just put your sinker just like you were goin' white trout fishin' or like you'd ground mullet fish, just a common little bottom-fishin' rig except with a bream hook and that little tiny piece of shrimp and you catch them croakers just one right behind the other. We catch our own bait every trip, first thing every morning.

I got a recreational boat for that, a little 18-foot Scout. I had to get something like that where I could rod-and-reel fish some because this net thing, with what we got left, has got so hard to where it's so separate and far apart to catch a few fish.

Who's still fishin' with nets? Now, some of them boys that are still younger, they work menhaden, catchin' the pogy fish. And a few of 'em are still tryin' to cast-net some mullet a little bit. But that's too seasonal to where you ain't gonna make it. There's no way you're gonna get a year-round living out of that.

I don't think it's over three licenses getting sold to haul seine boats over there now. If they are, I ain't aware of who they are and where they're fishin'.

Well, my older brother, Rufus Lee, has pretty much just quit. He ain't bought his net license in seven or eight years.

The O'Briens, they got rid of all their stuff. They're not doin' it.

Besides me, the only one I know now that is actually still tryin' is that Mark Lewis boy. He's kind of a new one that showed up on the scene. He come out there with a cast net and stopped, stared, and watched and watched us until he tried to learn. Then he ended up gettin' a net. He got a haul seine, but I think he's mostly selling menhaden. And whatever extra food fish he can catch.

And then Larry Ryan is doin' it to catch menhaden. He catches 'em and packs 'em to freeze for sport bait. And then he's got a little shrimp boat now. He's doin' live-bait shrimpin' and bull minnows, I think, and a few live croakers to the sport industry.

He's younger than I am. And he's lucky enough to be down there in a place close to the coast where he's got access to the people that buy that. I don't have that option. I live so far away.

I think he's buyin' that $500 purse-seine license. It's different than mine but you can only catch menhaden. You can't catch no food fish with that license, and I think they've moved 'em either a mile or two miles offshore.

It's real spotty on the fishin'. And with the limited area we got to fish now, it's just about got impossible to stay with it. They took so much area and so much away from us till we're about limited out, for the kind of fishin' what I do.

Then they can see the population explosion of the recreationals, the tourism explosion. And the restrictions on where you got left to fish, they know that sooner or later you can't keep operatin' because there's just not enough there to operate on. That's how they work it.

I've had them come up at some of those meetings and tell me, "Well, they know you're gonna get old, you ain't gonna be able to go, eventually."

They know you can't keep goin' forever. They know how you think. Or I know how they think because I've heard their comments at them meetings over there.

I'll tell you, it's hard to squeeze a few dollars out of what's left, no doubt.

My wife and I don't pay for TV. We watch these local channels and PBS. Unless I'm mistaken it was PBS and there was a thing showing these guys up on the Atlantic Coast. I'm trying to remember which state, but I think it was North Carolina or even north of that. I'm not sure—I am startin' to get this, what do they call it? CRS? You know, Can't Remember Stuff. But they were promotin' their bluefish.

They were gillnettin' but they said these aren't gillnets. I don't know how they're gettin' away with this—they must have a lot better fisheries people to deal with up there than what we got here.

They said, "We find our fish and we strike 'em." They was lettin' them work and they was settin' on them fish with gillnets but they was strikin'. They never set their nets straight out, they what we call round hauled—put a compass around 'em. Then they pick it up and they said they got a regular little market, but it was growin'.

I'm sure they were fishin' the ocean beach. We'd get bluefish when you could fish the islands, outside on the Gulf beaches. They would mostly be a pound or two pounds but every now and then you could catch 'em 3 to 5 pounds. And I've even seen some bigger than that.

Them guys on that show were talkin' about how they didn't used to ice 'em and take good care of 'em but now they put 'em in a slush. They salt their ice slush to where it's super cold and when they take 'em out of their net they drop 'em straight into that. Then they pack them on good ice and they said the quality is one hundred percent different than what it used to be 20 years ago. A beautiful product.

They said their main market was to people that like to smoke 'em. And I have ate 'em before smoked and they ain't bad.

I heard that they did finally appoint someone to the commercial fisherman position on the commission. Maybe an oyster farmer. But I haven't been over there to Biloxi. There's no reason to go over there—there's nothin' left.

They turned them into what they're calling an advisory commission anyway. When the commission had charge of it, we could go over there to that building, we could at least be there and fight and see what they was doin'.

Well, now, you can go over there at a public hearing and you can go blah blah all you want but it don't matter because what them representatives and all in Jackson's gonna do up there at that Capitol, is all that matters now. And what was said in that building in Biloxi don't mean nothin'. They ain't listenin'. They ain't even heard what you said.

So, in other words, all the change is happening in Jackson again now. There really ain't a way for us to fight no more. They seen we was puttin' up a pretty good fight with that commission, and they didn't like that so they turned it back over to where they could go up there and just do it by introducing them bills and make all the law changes up there.

Fresh and local from the wild. Martin Young's catch for the day included four species: a redfish or two, a few more spotted seatrout, even more puppy drum, and mostly striped mullet. *(Ryan Bradley photo)*

They took the political route where we couldn't fight 'em as good is what they done.

I was born into this fishing life. I didn't finish high school, I never got no diplomas, I had no college, and I mean I ain't an educated person. But I do credit myself with havin' good common sense.

So I went to tryin' to figure some of these things out and I found out where you could go and make these Freedom of Information Act requests where you can go over there and ask for stuff from them to where they had to give it to you.

For example, I get checked sometimes ten times a week. So I went over there and I made a FOIA request of that commission. I asked for all of the reports that they had on me, all of the calls that they had on me and all of the times that they had come to my boat and how many times that I had been checked.

They sent me back a letter that said, "We don't keep none o' these records you have requested."

They knowed that if they'd have said, "Well, we checked you a hundred times but we ain't checked so and so over here but five times," I'd have had a legal discrimination charge, that they were singling me out.

And I always thought all police and law enforcement kept records on every call, every check, but according to them they don't keep none o' them kind of records.

But, see, stuff like that, when they seen that dumb old fishermen, which I'm sure they call us—a bunch of dumb uneducated—when they went to seein' that we was tryin' to learn and do things that put them on the spot, I don't think they really liked that.

That incident at Cat in 2019? He was one of them people from over there around Bay St. Louis or the Pass Christian area. He was one of them lifetime net haters that was doin' that. And he passed away a couple years back.

This guy I'm talking about, he was buddies with all of 'em, the hardcore sports. A recreational man, for sure. He wasn't one of the main guys, but he was always in there complaining about it.

He was the one who was doin' the cussin'. He was saying all kinds of things. Tommy even made the comment, "My wife's in that boat. You need to watch your language." Oh man, he was lettin' us have it, yessir.

Then, what he did, he come up there and he was gonna hit my net. We was picking a set up and he run up there like he was gonna hit my net but it was shallow and he run aground and it stopped his boat before he could. He was talkin' about gettin' his knife out and cuttin' it and we told him, "Go ahead, do whatever."

Tommy O'Brien's wife, she was in his boat and Tommy was on the boat with us gettin' the net up. We all told him just go ahead and do what you're gonna do but the law is on their way.

So, that time we had called the law to come get them!

They had already checked us two or three times. They had got to where, a lot of days, they would be settin' there waitin' on us before we could get to Cat Island. People would be callin' the law, tellin' them we was goin' with the net that-a-way when we'd leave from Pascagoula.

We would run from Pascagoula to keep from goin' down there towin' them boats with them nets on the interstate and launchin' at them boat ramps down there. We would run all the way from Pascagoula to avoid harassment. Tryin' to keep from makin' any problems, you know?

Man, when you're hated, a lot of people go through life really never understandin' how that is, to be hated by your peers.

You can be the nicest person in the world, and they still hate you for what you do. You could be the biggest saint, you could be a preacher—it does not matter—and still be the most hated person on the planet just because of what you do to provide food for your family. What you do for your living is enough for 'em to hate you that bad. And that's how it's been. People don't realize how that really is.

CHAPTER FIVE

PARTING SHOTS

FUN

By 2023, the youngest Baby Boomers were pushing 60 while most of those born in the 1940s and 1950s were already retired. If enough of those retirees were spending their golden years on the water, it could explain why DMR biologists were noting a sharp increase in the number of times that anglers were going fishing even while license sales remained flat.

Whatever the reason, the increased fishing trips were resulting in more "fish coming out of the water, harvested."

Mississippi's recreational fishermen boated 8 million pounds of the coastal species in 2023, the most recent year that data were available before publication of this work. That was nearly double the 4.4 million pounds anglers caught in 1994.

In descending order of volume, those landings included spotted seatrout (3,267,654 pounds); red drum (1,242,507 pounds); sheepshead (813,651 pounds); sand seatrout (798,212 pounds); black drum (675,304 pounds); Gulf and southern kingfish (399,930 pounds); Gulf and southern flounder (373,782 pounds); striped mullet (220,944 pounds); gafftopsail catfish (101,758 pounds); Spanish mackerel (46,536 pounds); and pinfish (13,456 pounds).

FOOD

In 2022, DMR executive director Spraggins told the Commission on Marine Resources, "We want people, more than anything, to eat fresh-grown South Mississippi seafood. We want them to eat the food that is coming out of our Gulf, not out of some other country."

Seafood consumers could still find some locally caught saltwater fish, but it wasn't easy. Commercial landings in 2023 totaled 157,624 pounds, about one tenth of the 1994 catch.

In descending order of volume, those landings included striped mullet (54,083, $24,843); red drum (36,646, $119,173); spotted seatrout (33,132, $133,322); sheepshead (15,981, $18,343); black drum (6,645, $7,966); sand seatrout (2,180, $5,021); tripletail (2,065, $9,012); sea catfishes (2,046, $775); king whiting/kingfishes (845, $2,056); spot (350, $167); Spanish mackerel (323, $504); bluefish (197, $76); ladyfish (114, $68); and Florida pompano (46, $162).

The three biggest food fisheries—striped mullet, red drum and spotted seatrout—weighed a combined total of 123,861 pounds and had a dockside value of $277,338, about 96 percent of the entire inshore fishery's value.

Most of the reds and trout were taken with hook-and-line gear such as trotlines and rods and reels. Some mullets were taken in cast nets, and most of the remainder were seined. No gillnets or trammel nets—of any material—were registered for use.

COMPARISON WITH MISSISSIPPI'S "MIRROR IMAGE"

Back in the tumultuous 1990s, the commissioner of neighboring Alabama's Department of Conservation and Natural Resources had signaled that his agency would "stay out of the emotion" and avoid the "ban mania" that was sweeping other gulf states.

PARTING SHOTS

Alabama's managers did just that—they stuck to the facts, put the interests of the public and the resources first, and resolved the issue with a minimum of involvement by elected officials.

As a result, the Yellowhammer State's coastal fishermen continued to produce meaningful quantities of wild finfish. Although redfish and trout had been off limits as gamefish since the 1980s, netters targeted a couple dozen other species, 19 of them in significant amounts. From 1994 to 2006, the total annual landings of those 19 major species increased by 40 percent, from 3.5 million pounds to nearly 5 million pounds, partly in response to an upsurge in demand by Asian consumers for underutilized species like ladyfish and blue runners.

Recreational landings were also on the upswing. According to the National Marine Fisheries Service, Alabama's anglers in 1994 had made an estimated 2.8 million saltwater fishing trips and caught nearly 5.6 million pounds of their 15 most popular inshore species. By 2006, the anglers were making nearly twice as many trips—over 5.1 million—and catching one-third more fish—over 7.4 million pounds.

That both fisheries could grow while co-existing wasn't a good look for the Coastal Conservation Association's sport-only agenda. So, in 2007, the group's leaders rallied anglers for another attack.

This time, the sportsmen circumvented the scientifically oriented fishery managers at the state's Department of Conservation, smeared the commercial fishermen, and begged their representatives in the Legislature to ban the nets.

The political leaders stood with the seafood producers in 2007, but they soon wearied of the recreational anglers' incessant clamoring.

In the 2008 session, the Alabama Legislature made the fishermen an offer they couldn't refuse—either take some buyout bucks and go away or keep working until they were no longer able. Net licenses would not be transferable, and no new ones would be issued.

Landings by the finfish fishermen who opted to keep working naturally declined over the years as they aged out. Still, in 2023, those who remained brought in a respectable 2.3 million pounds of coastal finfish, about 15 times as much as Mississippi fishermen.

CHECKS AND BALANCES

While Mississippi's "conservationists" proclaimed recreational angling to be a higher use of the resource, from the fishes' perspective it made no difference whether a hundred sport fishermen each killed one, or a single gillnet fisherman killed 100—the impact on the population was identical.

In theory, the fish would benefit from a healthy tension between the user groups, with each checking the other. In practice, though, Mississippi's activist anglers focused as much on restricting the activities of the commercial fishermen as they did catching their own fish, And while the leaders of the state's commercial sector advocated for their constituents, they could not hope to out-lobby the CCA. Even so, they contributed viewpoints to the management narrative that would otherwise remain unsaid, such as:

- "There is a class going on right now over at the lab in Ocean Springs teaching people to fish. That's great, but it puts more pressure on the fish."

- "Everybody is getting a little smarter with social media. One person catches fish, and we've got everybody headed that way the next day. Things are changing and they are changing quickly. We've got to make management decisions with that in mind."

- "The problem is we've got too many hooks in the water. That's the whole problem; we are dancing all the way around it."

PARTING SHOTS

- "If you look at the changes in the way fishermen fish, it is almost to the point where, now, we ... have more fishermen than we have fish."

- "At some point in time, we are going to have to make management decisions that will actually recover the species."

- "We strongly encourage the commission to pursue true conservation measures such as implementing an annual total allowable catch [quota] in the recreational sector."

- "When you look at fisheries that are successfully managed most of them are managed under a quota system."

Mississippi's fishery managers were equitable enough to have maintained at least some commercial fishing for seatrout and even the hot-button redfish, which were both pricey enough to make hook-and-lining feasible. But when those quotas were filled, other options on the water were limited. As a result, participation in these highly valued fisheries was eroding, and the trajectory overall wasn't promising.

Meanwhile, who will oversee the recreational fishery in the future? A state agency funded by the sale of sport-fishing licenses? Coastal city councils? Sport-fishing guides? Boat and tackle dealers? Politicians?

As more people went fishing more often, the quantity of fish that each could retain had to be reduced. Ironically, as restrictions cut the number of fish that anglers could legally keep, they increased the number that were discarded. Not all those discards survived. Biologists estimated that about 3.5 percent of the relatively tough redfish, and 10 percent of the more fragile trout, and similar species, died after being discarded. Over millions of trips, those numbers added up.

While "catch and release" fishing elicits a warm and fuzzy feeling among those who don't know better, dead discards can become the dominant source of removals in a fishery even when they are virtually invisible.

STATE VS FEDERAL

By the 1990s, everyone with knowledge and experience in the field knew that sustaining abundant coastal fisheries required ongoing stewardship that included the aggressive acquisition of virtually any land that touched water, and shared allocations between all user groups.

Because the netting issue hadn't been approached in a calm manner back in the nineties, it was a little late in the game in the 2020s to be discussing equitable allocation of Mississippi's inshore stocks. But that didn't mean it couldn't be done, and for a model we needn't look any further than the waters just offshore of the state.

The red snapper is a reef fish that occurs primarily in federal waters, which extend seaward 200 miles, well beyond the state's three-mile ribbon. It's also another of the Gulf's hotly contested species, hammered by both sport and commercial fishermen until around 1990, when federal managers imposed a minimum size limit of 12 inches and a Gulf-wide quota of about 3 million pounds.

More than three decades later, the quota had grown five times larger and was shared almost equally between the recreational (49%) and commercial (51%) sectors.

Here's how the fishery was allocated in 2022:

After biologists set a Gulf-wide total allowable catch of 15.1 million pounds, the commercial sector's 51-percent allocation equated to 7.7 million pounds, which was distributed to the individual seafood producers according to their historical landings.

The recreational sector's fixed 49 percent of the TAC equated to 7.4

million pounds; of that, 42.3 percent, or 3.1 million pounds, was carved out for the charter-boat operators while private anglers got the remaining 4.3 million pounds.

The fixed and equitable allocation of the red snapper fishery seems mundane now but getting there wasn't easy. Commercial reef fishermen didn't use nets, they caught their fish with hooks, like the recreational anglers, but they used more hooks and were each permitted to bring in a lot more fish than the sportsmen could.

As such, the passions of anglers could have been as inflamed over red snapper as they had been over redfish, trout, and nets. Why then hadn't the offshore snapper fishery been regulated out of existence like the inshore fisheries?

Because it was managed federally, not by the state.

In 1976, the U.S. Congress created panels of stakeholders who were to manage each region's federal fisheries according to a set of ironclad standards including "allocation shall be fair and equitable to all fishermen, reasonably calculated to promote conservation, and carried out in such a manner that no particular individual, corporation, or other entity acquires as excessive share of such privileges." The process was overseen by the U.S. Department of Commerce, the lawmakers themselves, and the federal courts.

When the Mississippi Legislature created the Commission on Marine Resources in the mid-1990s, it modeled the panel's membership on that of the federal Gulf of Mexico Fishery Management Council and adopted virtually the same set of rules for state managers to follow. The difference was that federal lawmakers let the federal managers do their jobs while state legislators couldn't let their fishery managers alone.

Because nearshore fishing was more affordable than offshore fishing, there was a much larger pool of inshore anglers to lobby their local legislators, who were easier to intimidate than congresspeople from around the country.

Legislators are expected to act in the interest of all their constituents. In Mississippi, lawmakers interceded on the sport fishermen's behalf until there was next to nothing left.

In the preface to the Gulf Wars series, I wrote, "Even though I'd fished for years, commercially and recreationally, I learned much in writing this series—not just about the Gulf of Mexico's ecology and fisheries, but how mass movements are organized, and how man relates to himself, his fellows, and to nature. It's often not a pretty picture."

As we've seen, it's been a long, unhappy trip for Mississippi's fishermen since the "nonpartisan" philanthropies, nonprofits and mass media came into fishing in the early 1990s.

Under the circumstances, Magnolia State sportsmen didn't behave any worse than did anglers in other Gulf states. After all, it's hard to stay on the sidelines when your competition's being pummeled.

The upshot, as Pascagoula commercial fisherman Larry Ryan Jr. told the coastal commissioners, was that net fishermen weren't a "dying" breed: "We make it that way."

Appendix

TOOLS OF THE TRADE
A GLOSSARY OF SOME FISHING NETS

TABLE 1
MISSISSIPPI COMMERCIAL LANDINGS OF STRIPED MULLET
IN POUNDS AND DOLLARS 1950-2023

TABLE 2
MISSISSIPPI COMMERCIAL SPOTTED SEATROUT LANDINGS
IN POUNDS AND DOLLARS 1950-2023
&
RECREATIONAL SPOTTED SEATROUT LANDINGS IN POUNDS 1981-2023

TABLE 3
MISSISSIPPI COMMERCIAL RED DRUM LANDINGS
IN POUNDS AND DOLLARS 1950-2023
&
RECREATIONAL RED DRUM LANDINGS IN POUNDS 1981-2023

TABLE 4
NUMBER OF RECREATIONAL FISHING TRIPS IN GULF OF MEXICO
FOR ALL SPECIES 1981-2023

BIBLIOGRAPHY

TOOLS OF THE TRADE

A GLOSSARY OF SOME FISHING NETS

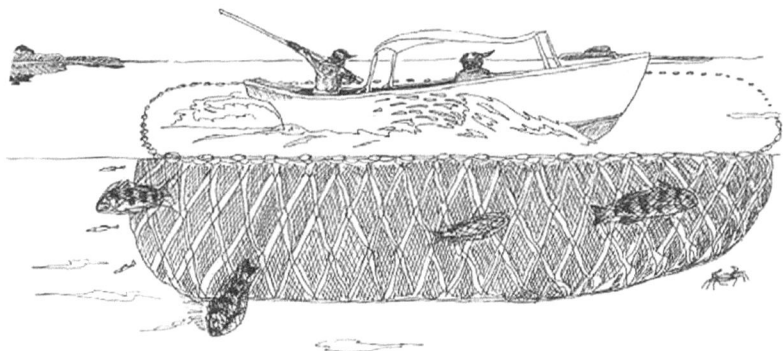

Fishermen herding black drum into trammel net.

TRAMMEL NET

The trammel net is actually three nets in one. Sandwiched between the two outer "walls," the inner net is fine-meshed and hung with plenty of slack. The walls are nets with extremely large meshes, up to 18 or 20 inches. When a small fish hits the trammel net, it may be gilled in the inner net; when a larger fish hits, it forces the small-meshed inner net to bulge through a large mesh on the opposite side, neatly bagging itself in a pouch.

The trammel net's construction enables fishermen to catch deep-bodied species lacking prominent gill covers—such as pompano, flounder and the freshwater buffalo—as well as a wider size-range of fish than they can with gillnets.

279

GILLNETS

Gillnets trap fish as they try to swim through their webbing. The nets are held vertically in the water by a floating corkline and a sinking leadline. If the fish is the correct size, its head will pass through a mesh but its body will be too large to follow. Unable to go forward, and prevented by its gill covers from backing out, the fish is held until removed by the fisherman.

Obviously, the larger the fish desired, the larger the mesh size must be.

The selectivity of the gillnet can prove frustrating for fishermen as smaller, yet marketable, fish pass through the webbing. Fish that are too large to enter the meshes often back off and escape, as well. Still, the gillnet's light weight and ease of handling and clearing more than compensate for the fish lost due to its discriminating construction.

The gillnet's selectivity is also useful for fishery managers, who can mandate minimum mesh sizes that won't catch fish until they've grown large enough to have spawned at least once, thus ensuring continuous production.

Gillnets may be fished actively or passively. A runaround gillnet, or "strike net," as it's called on the Gulf Coast, is fished actively: When the fisherman sights a school of fish, he throws over a weight or buoy that's attached to the net which is then pulled into the water as he circles the fish; after they hit the net, the fisherman hauls it back onboard.

A "set-net" is fished passively by anchoring it in a promising location, like a spider's web. Fishermen check their gear periodically to clear it of fish.

"Drifting" is sort of a combination of the two techniques: The fisherman runs out his net in a promising location, ties his boat to one end and stays with his gear while it fishes.

(Above) Fishermen hauling back strike net. *(Below)* A set-net with alligator gar.

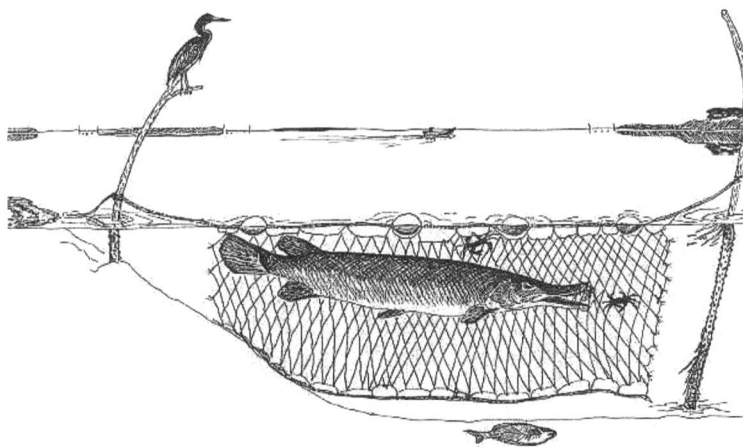

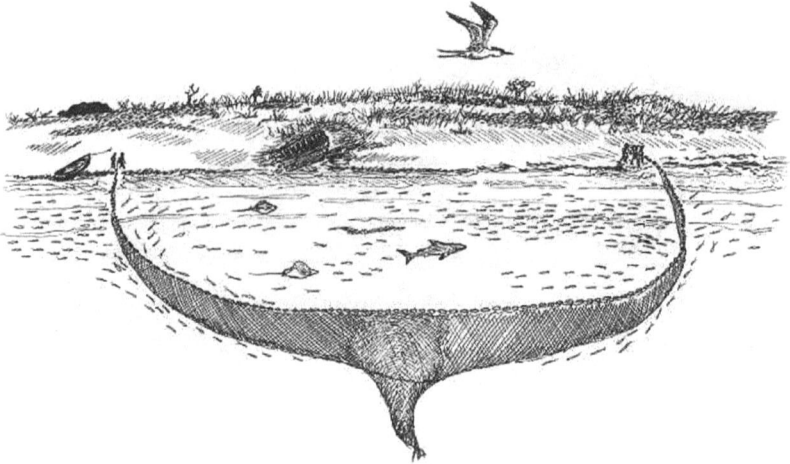

(*Above*) Fishermen preparing to drag haul seine back to shore.
(*Opposite*) Purse seine. Inset shows pursing cable through rings at bottom of net.

HAUL SEINE

Unlike gillnets and trammel nets, which entangle fish in their webbing, the haul seine is used to surround the catch with a wall of small-meshed webbing that is constructed of extremely coarse twine to withstand the wear and tear this gear is subjected to.

After the seine is set around a school of fish, it is slowly hauled back to the boat or shore. As it is "hardened up," the catch is gradually forced into a bag of webbing that's sewn into the net. Once it's confined within the bag, the catch can be scooped out with a dip net or gaffed individually; either method is more efficient than clearing single fish from an entangling gill or trammel net.

PURSE SEINE

The purse seine is a very large net used primarily to capture schools of fish found near the surface of deeper waters. Unlike trammel nets, haul seines and gillnets—the leadlines of which usually sink to the bottom to prevent fish from escaping beneath—the purse seine floats like a deep curtain of webbing. To prevent the catch from escaping by diving deeper, the net is "pursed" by heaving on a cable that's run through rings that are attached to its leadline. The net is then hauled in by power blocks or a net drum until the fish are herded to one end. They can then be dipped out with nets or suctioned aboard with fish pumps.

Tuna are harvested in the Pacific Ocean with purse seines that may weigh 50 tons. Smaller seines used to catch mackerel in the Atlantic may weigh 15 tons. Purse seines (along with gillnets and set-nets) are used in Alaska's salmon fishery but on the Gulf Coast use of purse seines is limited mostly to the menhaden fishery. Alaska's salmon seiners are limited to 58 feet in length.

283

TABLE 1

MISSISSIPPI **COMMERCIAL** LANDINGS OF STRIPED MULLET
IN POUNDS AND DOLLARS
1950-2023[1]

YEAR	POUNDS	DOCKSIDE VALUE
1950	242,800	$ 21,259
1951	431,200	$ 35,088
1952	420,100	$ 34,447
1953	288,200	$ 21,613
1954	405,000	$ 28,378
1955	439,200	$ 26,352
1956	438,900	$ 26,340
1957	357,100	$ 18,926
1958	549,300	$ 27,640
1959	561,600	$ 28,387
1960	395,100	$ 19,944
1961	400,400	$ 19,973
1962	506,900	$ 25,648
1963	382,200	$ 19,561
1964	249,500	$ 12,391
1965	240,800	$ 12,494
1966	636,400	$ 32,076
1967	1,704,800	$ 86,658
1968	947,000	$ 47,772
1969	387,900	$ 19,469
1970	161,900	$ 9,156
1971	176,700	$ 9,820
1972	221,000	$ 12,555
1973	482,200	$ 32,411
1974	451,600	$ 29,177

[1] *From National Marine Fisheries Service landings data.*

TABLE 1 *(continued)*

MISSISSIPPI COMMERCIAL LANDINGS OF STRIPED MULLET
IN POUNDS AND DOLLARS
1950-2023[1]

YEAR	POUNDS	DOCKSIDE VALUE
1975	284,500	$ 23,518
1976	840,900	$ 94,042
1977	948,600	$ 133,278
1978	1,487,200	$ 256,518
1979	1,482,000	$ 243,922
1980	1,989,980	$ 346,828
1981	1,300,950	$ 254,424
1982	459,500	$ 91,843
1983	800,010	$ 146,186
1984	763,976	$ 166,496
1985	46,353	$ 11,603
1986	1,125,496	$ 385,522
1987	585,468	$ 238,414
1988	700,161	$ 320,820
1989	253,152	$ 134,796
1990	803,253	$ 433,853
1991	439,334	$ 199,976
1992	474,334	$ 378,882
1993	247,078	$ 176,649
1994	781,300	$ 407,103
1995	615,142	$ 549,380
1996	842,186	$ 725,605
1997	473,688	$ 391,306
1998	319,214	$ 165,720
1999	522,274	$ 365,657

[1] *From National Marine Fisheries Service landings data.*

TABLE 1 *(continued)*

MISSISSIPPI COMMERCIAL LANDINGS OF STRIPED MULLET IN POUNDS AND DOLLARS 1950-2023[1]

YEAR	POUNDS	DOCKSIDE VALUE
2000	255,525	$ 166,705
2001	233,132	$ 113,561
2002	64,169	$ 21,696
2003	94,266	$ 33,518
2004	127,800	$ 53,899
2005	98,806	$ 37,545
2006	65,999	$ 23,241
2007	69,732	$ 35,043
2008	57,177	$ 32,403
2009	62,330	$ 29,993
2010	58,796	$ 30,552
2011	92,639	$ 56,448
2012	98,686	$ 62,938
2013	95,386	$ 61,039
2014	21,167	$ 13,216
2015	20,478	$ 11,026
2016	38,662	$ 20,107
2017	61,864	$ 33,797
2018	167,960	$ 68,038
2019	34,720	$ 17,457
2020	23,441	$ 11,728
2021	52,932	$ 13,593
2022	67,110	$ 13,800
2023	54,083	$ 24,843
Totals:	32,080,709 pounds	$8,286,062

[1] *From National Marine Fisheries Service landings data.*

TABLE 2

MISSISSIPPI **COMMERCIAL** LANDINGS OF SPOTTED SEATROUT
IN POUNDS AND DOLLARS, 1950-2023
&
RECREATIONAL LANDINGS OF SPOTTED SEATROUT IN POUNDS
1981-2023

YEAR	POUNDS	DOCKSIDE VALUE	RECREATIONAL[1]
1950	61,500	$ 15,375	–
1951	142,200	$ 31,284	–
1952	1,259,700	$ 69,075	–
1953	202,900	$ 56,812	–
1954	139,800	$ 37,746	–
1955	166,600	$ 41,650	–
1956	213,900	$ 53,470	–
1957	210,800	$ 52,700	–
1958	283,300	$ 70,833	–
1959	253,900	$ 63,475	–
1960	116,500	$ 29,130	–
1961	171,800	$ 43,924	–
1962	111,600	$ 27,895	–
1963	80,300	$ 20,074	–
1964	148,100	$ 30,522	–
1965	148,600	$ 37,138	–
1966	144,600	$ 36,101	–
1967	171,200	$ 42,747	–
1968	268,500	$ 66,546	–
1969	220,800	$ 54,595	–
1970	254,800	$ 63,247	–
1971	393,400	$ 98,397	–
1972	254,600	$ 67,918	–
1973	365,900	$114,014	–

TABLE 2 *(continued)*

MISSISSIPPI **COMMERCIAL** LANDINGS OF SPOTTED SEATROUT
IN POUNDS AND DOLLARS, 1950-2023
&
RECREATIONAL LANDINGS OF SPOTTED SEATROUT IN POUNDS
1981-2023

YEAR	POUNDS	DOCKSIDE VALUE	RECREATIONAL[1]
1974	294,700	$ 91,168	–
1975	262,800	$ 97,095	–
1976	177,500	$ 76,585	–
1977	147,000	$ 70,298	–
1978	105,420	$ 56,025	–
1979	109,480	$ 79,275	–
1980	27,245	$ 21,354	–
TOTALS: (1950—1980)	6,909,445	$1,716,468	
1981	8,980	$ 8,122	764,778
1982	16,820	$ 15,758	852,569
1983	54,060	$ 49,913	1,130,530
1984	55,003	$ 54,258	406,938
1985	47,426	$ 54,679	257,244
1986	38,035	$ 38,265	916,126
1987	57,304	$ 60,433	869,001
1988	65,584	$ 76,297	260,507
1989	77,616	$ 95,833	560,607
1990	30,442	$ 46,431	619,559
1991	31,295	$ 51,511	798,772
1992	31,564	$ 45,447	431,753
1993	51,362	$ 83,641	582,990
1994	73,034	$ 123,223	404,698

TABLE 2 *(continued)*

MISSISSIPPI **COMMERCIAL** LANDINGS OF SPOTTED SEATROUT
IN POUNDS AND DOLLARS, 1950–2023
&
RECREATIONAL LANDINGS OF SPOTTED SEATROUT IN POUNDS
1981–2023

YEAR	POUNDS	DOCKSIDE VALUE	RECREATIONAL[1]
1995[2]	72,158	$ 120,183	817,729
1996	43,589	$ 74,002	959,698
1997	41,456	$ 80,003	932,589
1998	42,562	$ 87,287	1,031,394
1999	50,510	$ 95,973	1,967,152
2000	45,501	$ 87,058	875,162
2001	43,146	$ 86,056	1,163,720
2002	32,483	$ 64,907	1,375,906
2003	25,940	$ 50,332	782,987
2004	30,469	$ 61,033	1,961,464
2005	18,850	$ 36,571	1,102,139
2006	23,404	$ 50,044	1,637,630
2007	28,098	$ 62,861	945,785
2008	33,533	$ 74,006	2,176,031
2009	52,615	$ 120,614	2,905,320
2010	41,534	$ 96,044	2,250,514
2011	38,675	$ 98,680	2,294,759
2012	61,099	$ 155,702	1,839,391
2013[3]	50,820	$ 127,977	2,244,643
2014	29,437	$ 88,133	1,535,624
2015	27,995	$ 82,804	2,869,641
2016	50,436	$ 161,241	5,246,130
2017	54,062	$ 237,013	2,551,405
2018	52,404	$ 162,046	2,813,449

TABLE 2 *(continued)*

MISSISSIPPI **COMMERCIAL** LANDINGS OF SPOTTED SEATROUT
IN POUNDS AND DOLLARS, 1950-2023
&
RECREATIONAL LANDINGS OF SPOTTED SEATROUT IN POUNDS
1981-2023

YEAR	POUNDS	DOCKSIDE VALUE	RECREATIONAL[1]
2019	36,953	$ 128,191	1,899,951
2020	20,814	$ 71,869	1,189,889
2021	23,795	$ 84,944	1,340,417
2022	28,252	$ 111,479	2,922,089
2023	33,132	$ 133,322	3,267,654
TOTALS:	1,772,247	$ 3,694,186	63,756,334
(1981—2023)	POUNDS		POUNDS

[1] *The federal government first began to annually compile recreational catch data in 1981. In 2017, NMFS updated the way it surveyed recreational anglers to more accurately reflect their actual catches. The agency then recalibrated landings for the years since 1981 and in most cases determined that anglers had in fact caught more fish than had previously been reported. Data in Table 2 may therefore not coincide with those cited in the text by biologists in the mid-1990s and other years prior to 2017.*

[2] *The Commission on Marine Resources placed a 40,000-pound quota on the spotted seatrout fishery.*

[3] *The Commission on Marine Resources increased the commercial trout quota to 50,000 pounds.*

TABLE 3

MISSISSIPPI **COMMERCIAL** LANDINGS OF RED DRUM
IN POUNDS AND DOLLARS, 1950-2023[1]
&
RECREATIONAL LANDINGS OF RED DRUM IN POUNDS
1981-2023[1]

YEAR	POUNDS	DOCKSIDE VALUE	RECREATIONAL
1950	51,600	$ 9,820	–
1951	31,400	$ 6,002	–
1952	41,300	$ 7,006	–
1953	61,700	$ 13,576	–
1954	60,700	$ 8,498	–
1955	56,800	$ 7,952	–
1956	71,300	$ 9,845	–
1957	53,600	$ 8,019	–
1958	65,000	$ 9,627	–
1959	71,400	$ 10,600	–
1960	38,900	$ 5,841	–
1961	52,900	$ 7,924	–
1962	76,000	$ 11,405	–
1963	59,000	$ 7,471	–
1964	50,100	$ 7,187	–
1965	32,700	$ 4,783	–
1966	36,900	$ 5,328	–
1967	95,800	$ 13,921	–
1968	214,600	$ 31,487	–
1969	99,600	$ 13,159	–
1970	70,300	$ 8,878	–
1971	58,800	$ 7,339	–
1972	55,700	$ 7,062	–

TABLE 3 *(continued)*

MISSISSIPPI **COMMERCIAL** LANDINGS OF RED DRUM
IN POUNDS AND DOLLARS, 1950-2023[1]
&
RECREATIONAL LANDINGS OF RED DRUM IN POUNDS
1981-2023[1]

YEAR	POUNDS	DOCKSIDE VALUE	RECREATIONAL
1973	85,700	$ 11,546	–
1974	88,600	$ 12,173	–
1975	71,500	$ 19,989	–
1976	95,200	$ 16,852	–
1977	155,300	$ 28,835	–
1978	658,000	$180,871	–
1979	194,380	$ 89,633	–
1980	20,430	$ 7,390	–
TOTALS: (1950—1980)	2,875,210 POUNDS	$590,019	
1981	66,955	$ 15,639	1,216,456
1982	40,600	$ 9,587	876,886
1983	24,200	$ 9,241	923,373
1984	23,660	$ 12,637	539,573
1985	27,423	$ 13,076	231,085
1986	126,352	$ 86,696	493,144
1987	53,059	$ 41,324	467,078
1988	41,109	$ 41,131	188,264
1989	139,775	$196,830	251,831
1990[2]	5,166	$ 7,576	259,338
1991	22,143	$ 28,529	217,269
1992	62,551	$ 85,653	189,712

TABLE 3 *(continued)*

MISSISSIPPI **COMMERCIAL** LANDINGS OF RED DRUM
IN POUNDS AND DOLLARS, 1950-2023[1]
&
RECREATIONAL LANDINGS OF RED DRUM IN POUNDS
1981-2023[1]

YEAR	POUNDS	DOCKSIDE VALUE	RECREATIONAL
1993	83,704	$ 106,216	458,842
1994	40,246	$ 48,927	448,295
1995	24,110	$ 30,446	827,159
1996	30,363	$ 38,042	1,199,863
1997	23,633	$ 30,603	1,266,562
1998	30,798	$ 51,123	1,119,200
1999	40,202	$ 65,167	779,923
2000	38,084	$ 57,977	695,394
2001	22,695	$ 35,540	588,507
2002	17,863	$ 24,916	1,214,244
2003	22,441	$ 30,443	816,557
2004	18,457	$ 24,938	1,041,015
2005	30,141	$ 38,684	682,284
2006	22,192	$ 29,474	1,020,848
2007	22,664	$ 33,340	692,036
2008	28,011	$ 42,329	1,172,861
2009	32,027	$ 50,432	1,023,111
2010	36,444	$ 64,619	977,967
2011	28,359	$ 57,893	845,774
2012	34,797	$ 69,279	1,372,678
2013	36,516	$ 74,510	3,114,427
2014[3]	43,243	$ 92,811	1,105,935
2015	61,492	$155,493	1,416,712

TABLE 3 *(continued)*

MISSISSIPPI **COMMERCIAL** LANDINGS OF RED DRUM IN POUNDS AND DOLLARS, 1950–2023[1]
&
RECREATIONAL LANDINGS OF RED DRUM IN POUNDS 1981–2023[1]

YEAR	POUNDS	DOCKSIDE VALUE	RECREATIONAL
2016	60,511	$ 149,809	1,599,600
2017	56,597	$ 139,759	1,861,951
2018	47,684	$ 115,502	1,893,808
2019	61,531	$ 154,178	3,583,285
2020	51,766	$ 132,975	1,112,675
2021	35,182	$ 91,542	1,453,382
2022	46,740	$ 143,953	1,074,754
2023	36,646	$ 119,173	1,242,507
TOTALS: (1981—2023)	1,798,132 POUNDS	$ 2,848,012	43,546,165 POUNDS

[1] *From National Marine Fisheries Service landings data.*

[2] *35,000-pound commercial quota implemented for red drum.*

[3] *Commercial red drum quota raised to 60,000 pounds.*

TABLE 4

NUMBER OF RECREATIONAL FISHING TRIPS
IN GULF OF MEXICO
ALL SPECIES 1981-2023[1]

YEAR	Florida[2]	Alabama	Mississippi	Louisiana	Texas	TOTAL[3]
1981	17,362,277	1,838,039	1,485,713	5,397,294	NA[4]	26,083,323
1982	22,314,511	2,019,598	1,882,086	6,243,856	NA	32,460,051
1983	27,339,191	2,021,204	2,050,852	6,720,797	NA	38,132,044
1984	29,246,713	1,957,765	1,954,953	6,511,012	NA	39,670,443
1985	28,433,066	2,038,196	1,941,006	6,571,295	NA	38,983,563
1986	23,896,697	2,078,320	1,976,017	6,123,108	NA	34,074,142
1987	24,911,619	1,991,245	1,921,081	5,836,312	NA	34,660,257
1988	29,611,091	2,031,182	2,074,563	5,877,407	NA	39,594,243
1989	31,591,138	2,065,211	2,172,414	5,892,249	NA	41,721,012
1990	29,059,881	2,199,281	2,611,196	6,470,610	NA	40,340,968
1991	36,269,447	2,503,643	2,818,480	6,488,761	NA	48,080,331
1992	29,880,423	2,523,414	2,642,645	6,740,854	NA	41,787,336
1993	34,266,205	2,590,633	2,696,126	7,148,479	NA	46,701,443
1994	33,534,745	2,798,021	2,728,903	7,230,755	NA	46,292,424
1995	31,337,372	2,821,251	2,749,208	7,266,614	NA	44,174,445
1996	30,151,619	2,905,562	2,865,138	7,824,715	NA	43,747,034
1997	30,184,122	3,251,103	3,157,758	8,116,784	NA	44,709,767
1998	32,465,528	3,274,381	3,291,403	8,262,188	NA	47,293,500
1999	37,243,772	3,459,088	3,585,927	8,739,158	NA	53,027,945
2000	38,312,644	3,744,313	3,791,218	9,835,541	NA	55,683,716
2001	40,356,323	4,018,024	3,904,563	10,047,817	NA	58,326,727
2002	36,897,038	4,109,359	3,743,348	9,366,820	NA	54,116,565
2003	40,094,000	4,397,398	3,760,258	9,770,130	NA	58,021,786
2004	48,059,682	4,742,003	3,718,305	9,111,752	NA	65,631,742
2005	41,879,648	4,376,978	3,259,936	8,548,931	NA	58,065,493

TABLE 4 *(continued)*

NUMBER OF RECREATIONAL FISHING TRIPS IN GULF OF MEXICO ALL SPECIES 1981–2023[1]

YEAR	Florida[2]	Alabama	Mississippi	Louisiana	Texas	TOTAL[3]
2006	37,528,052	5,110,383	3,857,950	8,029,154	NA	54,525,539
2007	40,137,729	5,219,235	4,031,282	8,952,616	NA	58,340,862
2008	40,894,492	4,977,530	4,353,513	9,805,845	NA	60,031,380
2009	36,207,467	5,441,596	4,573,165	10,515,738	NA	56,737,966
2010	39,446,328	5,329,293	4,509,351	11,222,564	NA	60,507,536
2011	40,063,360	5,737,821	4,503,301	11,453,646	NA	61,758,128
2012	44,997,654	6,150,613	4,492,747	10,889,486	NA	66,530,500
2013	46,293,290	6,768,525	4,341,700	10,770,452	NA	68,173,967
2014	38,625,282	6,481,789	4,311,510	NA	NA	49,418,581[5]
2015	35,730,006	6,829,718	4,593,570	NA	NA	47,153,294[5]
2016	38,936,416	7,319,601	4,717,914	NA	NA	50,973,931[5]
2017	41,840,176	8,493,330	4,851,754	NA	NA	55,185,260[5]
2018	40,996,400	6,680,646	4,554,960	NA	NA	52,232,006[5]
2019	35,644,843	6,676,884	4,226,768	NA	NA	46,548,495[5]
2020	42,197,504	6,623,029	4,297,896	NA	NA	53,118,429[5]
2021	37,559,775	6,815,020	4,774,808	NA	NA	49,149,603[5]
2022	40,293,496	7,424,334	4,714,059	NA	NA	52,431,889[5]
2023	45,017,887	8,177,688	4,436,777	NA	NA	57,632,352[5]
TOTALS:	1,527,108,909	188,012,378	148,921,949	267,784,740[6]	NA	2,131,827,976[7]

[1] Data from the NMFS Marine Recreational Fishing Survey. Includes trips in both inshore and offshore waters for **all** species, not just redfish and trout.
[2] West coast only
[3] Excludes trips in Texas for all years and trips in Louisiana for 2014-2023
[4] Not available from NMFS
[5] Includes trips for Mississippi, Alabama and west coast of Florida only
[6] Excludes trips during years 2014-2023
[7] Excludes all trips in Texas and trips in Louisiana for 2014-2023

BIBLIOGRAPHY

COASTAL EROSION

American Shrimp Processors Association. In Memoriam: Richard Gollott, a Pillar of the Wild-Caught Gulf Shrimp Industry: American Shrimp Processors Association. www.americanshrimp.com. May 11, 2021.

Barnett, Jane. "Sport Group Uses Conservation to Fight Commercial Fishermen. *National Fisherman*, Vol. 68, No. 6, 1987.

Bellande, Ray L. Casinos, Gambling, Liquor and Vice: Biloxi's Gaming and Casino History Time Line. Biloxi Historical Society. www.biloxihistoricalsociety.org/casinos-gambling-liquor-and-vice

Biloxi Waterfront Master Plan. Sponsored by the City of Biloxi, Greater Biloxi Economic Development Foundation and the Biloxi Port Commission, February 1985.

Boudreaux, Edmond. "The Seafood Capital of the World: Biloxi's Maritime History." The History Press, Charleston, S.C., 2011.

Carter, Hodding and Anthony Ragusin. 1951. "Gulf Coast Country." Duell, Sloan & Pearce, New York, N.Y.

Dawkins, Hunter. Commission Denies Ban on Net Fishing Outside Cat Island. *Gazebo Gazette*, Pass Christian, Miss. September 18, 2018.

_____. CMR Adopts Resolution Opposing Cat Island Legislative Bills. *Gazebo Gazette*, Pass Christian, Miss. April 7, 2020.

Duff, John A. & Harrison, William C. The Law, Policy, and Politics of Gillnet Restrictions in State Waters of the Gulf of Mexico. *St. Thomas Law Review*, Vol 9. Winter 1997. pp. 414-416.

Easley, J.E., V.K. Smith, M.K. Wohlgenant and W.N. Thurman. Allocating Recreational-Commercial Fishery Harvests: Literature Reviews and Preliminary Work Toward Modeling the Issue. Gulf and South Atlantic Fisheries Development Foundation, Inc., Tampa, Fla., March 1989.

Freese, Curtis H., Editor. "Harvesting Wild Species: Implications for Biodiversity Conservation." The Johns Hopkins University Press, Baltimore. 1997.

Fritchey, Robert. Biloxi: For more than 100 years, the seafood industry has played a pivotal role in the fortunes of this Mississippi coastal city. *National Fisherman* Yearbook, 1989.

_____. Anatomy of a well-oiled campaign to Ban the Nets! *National Fisherman*, December 1994.

_____. Southern Netters Fight Campaigns to Ban Their Gear. *National Fisherman*, July 1995.

_____. "Missing Redfish: The Blackened History of a Gulf Coast Icon." New Moon Press, Golden Meadow, La., 2017.

_____. "Let the Good Times Roll: Louisiana Cashes in its Chips with the 1995 Net Ban." New Moon Press, Golden Meadow, La., 2017.

_____. "A Different Breed of Cat: How Alabama Resisted the Net Bans that Swept the Gulf States in the 1990s only to Follow Suit in the New Century." New Moon Press, Golden Meadow, La., 2020.

_____. "Kiss Your Seafood Goodbye? Saturating Media Campaign Set the Stage for 1990s Gulf Wars." September 25, 2022. www.newmoonpress/fish school/kiss your seafood goodbye

BIBLIOGRAPHY

Galtsoff, Paul S., editor. "Gulf of Mexico: Its Origin, Waters, and Marine Life." Fishery Bulletin of the Fish and Wildlife Service, Vol. 55, U.S. Government Printing Office, Washington, D.C., 1954.

Greater Biloxi Economic Development Foundation. General Market Analysis, City of Biloxi, Miss., February 1988.

Grover, J.H. (ed.) Allocation of Fishery Resources. UN Food and Agricultural Organization. Auburn (Ala.) University, April 1980.

Gunter, Gordon. 1952. Historical changes in the Mississippi River and the adjacent marine environment.
Publications of the Institute of Marine Science. 2 (2): 119-139.

_____. 1953. The relationship of the Bonnet Carré Spillway to oyster beds in Mississippi Sound and the "Louisiana Marsh," with a report on the 1950 opening.
Publications of the Institute of Marine Science. 3 (1): 17-71.

_____. 1955. Mortality of oysters and abundance of certain associates as related to salinity. Ecology 36 (4): 601-605.

_____. 1963. The Fertile Fisheries Crescent. Journal of the Mississippi Academy of Sciences, Vol. 9, pp. 286–90.

Gutierrez, C. Paige. "The Cultural Legacy of Biloxi's Seafood Industry." The City of Biloxi, Miss., 1988.

Haines, Francis. "The Buffalo: The Story of American Bison and Their Hunters from Prehistoric Times to the Present." University of Oklahoma Press, Norman and London. 1970, 1995.

Helvarg, David. Untangling the Problem of Ocean Plastic: Mary Crowley captains the search for discarded fishing nets. December 27, 2019. https://www.Untangling the Problem of Ocean Plastic/Sierra Club

Hill, Samuel. Mississippi fishermen narrowly dodge Cat Island net ban: CCA-backed proposal fails in tie vote. *National Fisherman*, December 2018, pp. 17-18.

Hoffer, Eric. "The True Believer: Thoughts on the Nature of Mass Movements." HarperCollins Publishers, Inc., New York, 1951.

Husley, Val. "Maritime Biloxi." Images of America Series, Arcadia Publishing, Charleston, S.C., 2000.

Leard, Richard, et al. The Striped Mullet Fishery of the Gulf of Mexico, United States: A Regional Management Plan. Gulf States Marine Fisheries Commission, Ocean Springs, Miss. December 1995.

Lee, Anita. "There's Going to Be No Fishing." Can Mississippi Marshes Be Saved from Sea Level Rise? *The News Observer*. September 28, 2021.

Lorio, Wendell, et al. "The Relative Impact of Netting and Sport Fishing on Economically Important Estuarine Species." Mississippi-Alabama Sea Grant Publication-79-025, October 1980.

Lucas, Richard. Horn and Clark Families Receive Signal Honor for Excellence in the Seafood Industry. July 18, 2024. www.ourmshome.com/horn-and-clark-families-receive-signal-honor-for--excellence-in-the-seafood-industry/

Lyles, Charles H. "A Study of the Commercial Finfish in Coastal Louisiana." Louisiana Department of Wildlife & Fisheries Technical Bulletin, 1979.

BIBLIOGRAPHY

Lynch, Allan. Ghost Gear Haunts the World's Oceans. September 30, 2016. https://www.Ghost Gear Haunts the World's Oceans/Sierra Club.

Maghan, Bruce W. "The Mississippi Oyster Industry." United States Department of the Interior, Fish and Wildlife Service, Bureau of Commercial Fisheries, Fishery Leaflet 607. Washington, D.C., December 1967.

Matthiessen, Peter. "Men's Lives." Random House, Inc. New York, 1986.

McClane, A.J., editor. "McClane's New Standard Fishing Encyclopedia and International Angling Guide." Holt, Rinehart and Winston, New York, 1974.

Mississippi Department of Marine Resources. Mississippi Gulf Coast National Heritage Area Management Plan, Comprehensive Resource Management Planning, Biloxi, Miss., December 2005. https://z0sqrs-a.akamaihd.net/268-mgcnha/MSHeritageArea/MGCNHA-ManagementPlan-final.pdf

The Mississippi Legislature. The Joint Committee on Performance Evaluation and Expenditure Review. Report #344. A Review of the Adequacy of the Mississippi Gaming Commission's Regulation of Legalized Gambling in Mississippi. September 11, 1996.

Morton, Robert A. Historical Changes in the Mississippi-Alabama Barrier Islands and the Roles of Extreme Storms, Sea Level, and Human Activities. U.S. Geological Survey. June 2007.

National Marine Fisheries Service. Commercial Fishing Landings Data: www.st.nmfs.noaa.gov/commercial-fisheries/commercial-landings/annual-landings/index

_____. Recreational Fishing Landings Data: www.st.nmfs.noaa.gov
/recreational-fisheries/data-and-documentation/queries/index

_____. History of Management of Gulf of America Red Snapper. NMFS.
www.fisheries.noaa.gov/southeast/sustainable-fisheries/history-
management-gulf-america-red-snapper.

Otway, N.M. and Macbeth, W.G. Physical effects of hauling on seagrass
beds. Fisheries Research & Development Corporation Project No.
95/149 & 96/286, August 1999.

Ownby, Ted. "Prohibition." July 11, 2017, University of Mississippi.
www.mississippiencyclopedia.org/entries/prohibition/

Perkins, Garey B., Ph.D. "Economic Value of the Seafood Processing
Industry in Mississippi, 1978." Food and Fiber Center, Mississippi
Cooperative Extension Service, MSU, November 1979.

Posadas, Benedict C. January 1996. Economic Impact of Dockside Gaming
on the Commercial Seafood Industry in Coastal Mississippi. Mississippi-
Alabama Sea Grant Program Publication No. 95-015, Mississippi State
University Coastal Research and Extension Center, Biloxi, Miss.

Powell, Murella Hebert. 1998. Biloxi, Queen City of the Gulf Coast. In
"History, Art, and Culture of the Mississippi Gulf Coast," edited by
Lawrence A. Klein, Mary Landry and Joe E. Seward, pp. 130-160.
Marine Resources and History of the Mississippi Gulf Coast. Volume I.
Mississippi Department of Marine Resources, Biloxi, Miss.

Rainer, David. Seatrout, Flounder Limits Change August 1.
Outdoor Alabama: Alabama Department of Conservation and
Natural Resources. July 2, 2019.

BIBLIOGRAPHY

Ruello, N.V. and Henry, G.W. Conflict Between Commercial and Amateur Fishermen." *Australian Fisheries*. March 1977.

Schmidt, Aimee. Down Around Biloxi: An Overview of Ethnic and Occupational Identity in a Coastal Town. Chapter in "Ethnic Heritage in Mississippi: The Twentieth Century," edited by Shana Walton & Barbara Carpenter. Oxford Academic Books, University Press of Mississippi, 2012.

Sheffield, David A. and Nicovich, Darnell L. "When Biloxi Was the Seafood Capital of the World." The City of Biloxi, Miss., 1979.

Stephens, Deanne. "The Mississippi Gulf Coast Seafood Industry: A People's History." (America Third Coast Series). University Press of Mississippi, 2021.

U.S. Environmental Protection Agency. The Ecological Condition of Estuaries in the Gulf of Mexico. U.S. EPA, Office of Research & Development, Washington, D.C. 20460. EPA 620-R-98-004, July 1999.
www.epa.gov/sites/production/files/2015--/documents /ecocondestuariesgom_print.pdf

Works Progress Administration, Federal Writers' Project in Mississippi (WPA). 1938. "Mississippi, A Guide to the Magnolia State." The Viking Press, New York.

WHAT'S ON YOUR SHELF?

OTHER BOOKS BY NEW MOON PRESS

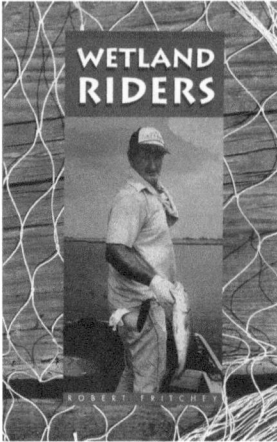

WETLAND RIDERS
1994

Like salmon in Alaska, the red drum was an essential ingredient of Louisiana's coastal culture. The "redfish" didn't run up rivers, it lived in the marsh where commercial fishermen had long harvested them for restaurants and markets. After the state interrupted the sale of the species, "for three years," fisherman Robert Fritchey hoped to win public support for reopening the fishery. He moved to the New Orleans French Quarter and investigated the origins of the movement that was monopolizing fisheries not just in Louisiana but around the country. His groundbreaking "Wetland Riders" is the result. Now a timely classic, this book is an essential primer for understanding how entitled anglers and the recreational fishing industry are displacing America's seafood production.

MISSING REDFISH
2017

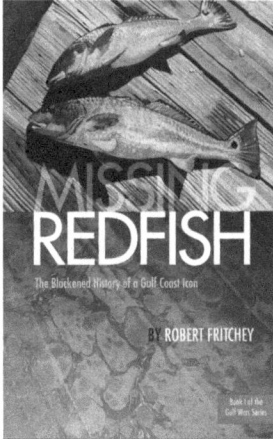

From Missing Redfish:

"As a mythical sea creature, the redfish ranks with Jonah's regurgitating behemoth and the vengeful white whale of Melville, the difference being that the red drum really does exist, and that the myths surrounding this bitterly contested fish originated not in great literature but in the slick campaign rhetoric of privileged anglers from the land of tall tales, Texas."

Commercial fishermen traditionally harvested juvenile redfish from inshore waters, but when New Orleans chef Paul Prudhomme created blackened redfish, some fishermen targeted the mature fish that lived offshore. When researchers sampled their catches, they learned that the brood stock had been declining for years before Chef Paul blackened his first fish. The cause? Both sport and commercial fishermen had been allowed to catch too many young fish. In his prescient "Missing Redfish," author Robert Fritchey suggests that sampling the offshore population could once again reveal trouble in the inshore population. This time there wouldn't be a commercial fishery to blame.

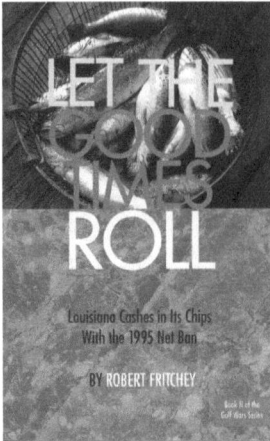

LET THE
GOOD TIMES ROLL
2017

South Louisiana's commercial fishermen netted tens of millions of pounds of coastal fish including red drum, spotted seatrout, flounder, black drum, striped mullet and the sumptuous Florida pompano. Day boats left the dock in the morning and returned in the evening with a freshly chilled catch that was treasured by New Orleans chefs and other consumers. The net fishery was sustainable and, to make it even more so, fishermen tried to boost efforts to retard the erosion of the state's coastal wetlands. When national environmental nonprofits and their funders began to smear commercial fishing in the early 1990s, sport fishermen sprang into action. In a state that called itself the Sportsman's Paradise, they ditched scientific resource management and begged their politicians to "Ban the Nets!" In 1995, they did. Like turning off a light, Louisiana traded food production for tourism. Let the Good Times Roll!

Louisiana Cashes in Its Chips
With the 1995 Net Ban

BY ROBERT FRITCHEY

Book II of the
Gulf Wars Series

THE GULF WARS SERIES
BOOK THREE

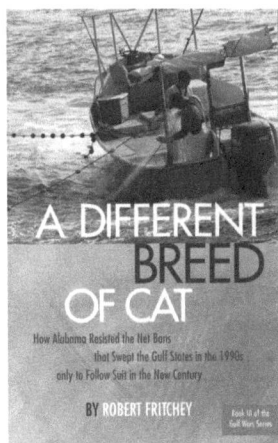

A DIFFERENT BREED OF CAT
2020

A DIFFERENT
BREED
OF CAT

How Alabama Resisted the Net Bans
that Swept the Gulf States in the 1990s
only to Follow Suit in the New Century

BY ROBERT FRITCHEY   Book 14 of the
Gulf Wars Series

After recreational fishing interests convinced Florida voters to outlaw most commercial fishing nets in the November 1994 election, sportsmen rose in unison across the Gulf of Mexico. Marshalled by the Texas-based Coastal Conservation Association, recreational anglers in Alabama, Mississippi and Louisiana demanded that their own states Ban the Nets! With a national "Global Fish Crisis" media campaign as backdrop, hysteria over a threatened invasion by out-of-work commercial fishermen from the Sunshine State presented sportsmen in the three central-Gulf states with a once-in-a-lifetime opportunity, and they weren't about to take "no" for an answer.

The battles raged over the spring and summer of 1995. Fishery managers in Louisiana and Mississippi buckled under the sportsmen's incessant clamoring, and the future looked bleak for Alabama's family fishermen as well.

Then the cream rose to the top.

Virtually every one of Alabama's institutions, including its natural resource management agency, media, legislators, even the governor, chose facts over emotion and helped preserve the public's sustainable five-million-pound fishery. In so doing, they made this characteristically hidebound state appear downright progressive—*A Different Breed of Cat.*